GERMAN MILITARY EFFECTIVENESS

GERMAN MILITARY EFFECTIVENESS

by Williamson Murray

 The Nautical & Aviation Publishing Company Of America
Baltimore, Maryland

Library of Congress Catalog Card Number 91-23601

ISBN: 1-877853-11-9

Printed in the United States of America

Library of Congress Cataloging in Publication Data

German military effectiveness/by Williamson Murray.
p. cm.
Includes bibliographical references and index.
ISBN 1-877853-11-9: $29.95
1. Germany—Military policy. 2. Germany—History, Military—20th century. 3. World War, 1939-1945—Germany. 4. World War, 1939-1945—Campaigns—Eastern. 5. World War, 1939-1945—Campaigns—Western. 6. World War, 1939-1945—Aerial operations, German. 7. Germany, Luftwaffe—History—World War, 1939-1945. I. Title.
UA710.M848   1991
355' .0332' 43—dc20
91-23601   CIP

DEDICATION

To Donald Kagan, who kept me in academia

CONTENTS

LIST OF TABLES

ACKNOWLEDGEMENTS

This book represents a collection of articles on the German military published over the period of a decade. A number of individuals read these articles and provided invaluable criticism and helpful suggestions. Among many others, my friends MacGregor Knox, Allan Millett, Richard Kohn, and Jürgen Förster read all or several of these pieces. Steven Glick and Russell Hart provided invaluable help in editing and preparing this manuscript from the raw material of the articles. I would also like to thank William T. and Dorothy P. Gilbert for their many kindnesses.

I must also thank the journals and publishers who originally selected these articles for publication. Seven Locks Publishers granted permission to republish "The German Response to Victory in Poland: A Case Study in Professionalism," *Armed Forces and Society* (Winter 1981); Frank Cass & Co. Ltd. allowed republication of the following articles: "The Luftwaffe before the Second World War: A Mission, A Strategy?," *Journal of Strategic Studies* (September 1981), "A Tale of Two Doctrines: The Luftwaffe's 'Conduct of the Air War' and the USAF's Manual 1-1," *Journal of Strategic Studies* (September, 1983), and "Clausewitz: What the Germans Got Right," in *Clausewitz and Modern Strategy*, ed. Michael Handel (London, 1986). The Royal United Services Institute published in its journal "Force Structure, Blitzkrieg Strategy, and Economic Difficulties: Nazi Grand Strategy in the 1930s," (April 1983). "German Air Power and the Munich Crisis" appeared in *War and Society*, vol. II, ed. By Brian Bond and Ian Roy (London, 1977). Two of the articles in this volume are the result of work supported by the Office of Air Force History and are scheduled for publication in the near future: "The Luftwaffe and Close Air Support, 1939-41" will appear in summer 1991 in Case Studies in Close Air Support, ed. by Richard Kohn; and "The Luftwaffe against Poland and the West" will appear in Case Studies in Air Superiority, also ed. by Richard Kohn.

As in all volumes of this sort, many people contributed to its good points; the faults are mine alone.

INTRODUCTION

Over the past fifteen years, in addition to publishing a number of books, I have worked my ideas out in articles that have appeared in a wide variety of publications. Those articles, in particular those on the German military, have extended my views on a number of issues and represent a body of work that has carried my work well beyond my books. Consequently, working with my publisher Jan Snouck-Hurgronje, I conceived of bringing these articles together in a collection on the German military and strategic situation.

To provide a framework, I have written an introductory essay on German military effectiveness from 1914 to 1945; that essay draws heavily on the intellectual conception behind the three volume study that Allan Millett and I edited, *Military Effectiveness* (London, 1988), as well as my own research and reading on the German military in this period. Nevertheless, no introductory essay could provide a fully coherent framework for any collection of articles. Consequently, the author must warn the reader that these articles address a number of disparate issues dealing with the German army and air force and that in some cases there are repetitions between the articles. I have attempted to minimize this and in such cases they reflect efforts to bolster very different arguments. Still the author asks the reader's forbearance; this book is a collection of articles written to stand independently, not as a coherent whole depicting the chronological history of the German military. Hopefully each of these efforts will extend the reader's understanding of the history of the Luftwaffe and the German army.

Williamson Murray
Secretary of the Navy Fellow
Naval War College
Newport, RI
November 1991

CHAPTER
ONE

THE PROBLEM OF GERMAN
MILITARY EFFECTIVENESS 1900–1945

The first half of our century was the time of the German problem. Emerging from centuries as "the Germanies," Bismarck's Second Reich had by 1900 established itself as the dominant continental power—one with every expectation and prospect of becoming a true world power. On the weight of its industrial development, the educational level of the population, cultural and national discipline, and geographic location, Germany seemed on the verge of emerging as a superpower, overshadowing her European rivals, including Great Britain, and perhaps challenging the United States. But this did not come to pass. In two great world wars Germany caused her own downfall, and in the resulting disaster drowned herself, her neighbors, and much of the world in a sea of blood.

The nature of that European catastrophe had much to do with Germany's attitude toward her military services, the peculiar cultural and intellectual forces that shaped those institutions, and a combination of battlefield brilliance mixed with a myopic strategic vision. That last combination proved deadly and in both 1914 and 1939 Germany embarked on conflicts that offered only the slightest chance of victory. As early as the failure before Paris in September 1914 and the disaster before Moscow in December 1941, Germany reached turning points which made defeat inevitable. Yet the tactical and operational competence of German military forces insured that the final defeat in those two world wars would not come for four more years and at a horrendous cost to all concerned.

By definition, military institutions in the western world exist to protect their nations from internal as well as external threats. The nature of the Second Reich's birth, built on the victories over Austria in 1866 and France in 1870, insured that its military institutions acquired a unique reputation and position in German society: a true irony given that the real reason for success in 1866 and 1870 lay in the masterful strategic and political execution of policy by Bismarck. In fact, the generals, for all their tactical and operational expertise, had again and again attempted to thwart Bismarck's vision in the pursuit of ephemeral battlefield success. This had been the case after Königgrätz, when Moltke sought to pursue the beaten Austrians to Vienna—risking bringing the other European powers into the conflict and endangering the strategic and political possibilities that Bismarck's diplomacy had created.

But the Prussian-Germans saw little of this in the euphoria over their revenge at Sedan-Metz for the 1806 disaster at Jena- Auerstadt. The economic successes of succeeding years helped to cover up the serious political and constitutional weaknesses of the Bismarckian Reich—especially after Wilhelm II had ascended to the throne. One of the great historiographical arguments since 1945 has has revolved around the degree of continuity between the institutions and policies of the Third Reich and its predecessors, the Weimar Republic and the Second Reich. While we will leave most continuities within German history for others to discuss, those within the military institutions of the German state are a basic theme of this volume; and it is only in understanding them that one can recognize the military's contribution to the breakup of the Bismarckian state.

Geography and history have played a crucial role in the approach which the German military took in tackling the problems that they confronted. The Reich's position at the center of Europe brought the German nation important advantages. Located in the heart of Europe, Germany was a cross-roads through which not only economic but cultural and intellectual currents flowed back and forth across the Continent.

But this position also had disadvantages. While the mountains to the south provide some shielding, its position on the north German plain opened the Reich to invasion from both east and west. Moreover, historical divisions dating from the Middle Ages made German territory forever tempting to foreign conquerors and states. By the eighteenth century, armies from virtually all of the major European powers, Sweden, France, England, and Russia, engaged in fighting on German territory. Consequently, due to geographic position alone, the German military have had to think of war on their immediate doorstep. The result has been an emphasis on operations and battlefield performance, while their counterparts in Britain and particularly the United States have had to think in terms of getting to the war.

At the same time, history has also had a considerable impact on the perceptions of the German officer corps. Above all the catastrophe of Jena-Auerstadt, where in fall 1806 Napoleon's brilliance and the Grand Army's excellence destroyed the Prussian army and state in one day, reinforced the importance of battlefield performance for the Prussian officer corps.[1] As a result, the Prussians and their successors in the German army, particularly in the general staff, undertook the study of operations and tactics with a seriousness that did not occur in other armies. Unfortunately for their nation and Europe, that study concentrated on operational and tactical concerns rather than on strategy. Not surprisingly, Napoleon Bonaparte lay at the heart of that study and, like the great master, the Prussians confused battlefield success with strategic wisdom.[2]

Political Effectiveness

Military institutions by nature are highly structured organizations that must interact with each other in the joint arena and with the political authorities who determine national policy. What is indeed astonishing about the history of the German military in the first half of this century is the poor organization of their external and internal bureaucratic structures. Not only did they work at cross purposes with each other, but their internal structures suggest a surprising unwillingness to think through basic administrative problems. Under the Bismarckian Reich three independent army authorities, the Prussian war ministry, the military cabinet, and the general staff all reported directly to the emperor.[3] How little cooperation occurred among the three is suggested by the fact that the general staff did not deem it necessary to inform the war ministry about the scale of the Schlieffen Plan until 1912.[4] The navy was no better organized. Nor did the army and navy regularly talk to each other or coordinate plans or wartime objectives. To add to the bureaucratic nightmare before World War I, Wilhelm II gave the uncoordinated right of access to his person to no less than forty army and eight naval officers.[5]

Consequently, an almost complete absence of policy and strategy accompanied Germany as she stumbled into war in 1914 with one unrealistic war plan and no mechanism for coordination of military force, diplomacy, and strategy. No larger body such as the Committee of Imperial Defense in Britain existed; as the last prewar chancellor, Bethmann Hollweg, lamented after the war, "there never took place during my entire period in office a sort of war council at which politics were brought into the military for or against 'considerations.'"[6] Without a political system capable of making sophisticated judgments, the German army seized *de facto* control of the war's direction after August 1914. This only compounded the kinds of mistakes that had dragged the German government into the quagmire in summer 1914; the decision to embark on unrestricted submarine warfare is a case in point.

The Weimar Republic's constitution sought to bring some order to the administrative chaos that had characterized the Bismarckian Empire. According to the new constitutional organization, a defense minister reported directly to the chancellor; as such, the minister was responsible for defense matters before the Reichstag with the army and navy working under his supervision. This system of cabinet responsibility for defense matters represented an organizational system similar to that used by most other western European nations. Unfortunately for the Weimar Republic, the system did not function as planned. The first defense minister, Otto Gessler, deferred to the military and in effect became their spokesman rather than a directing force to manage the new republic's military institutions. As before the war, little coordination existed between army and navy. Only during the Republic's last years under Wilhelm Groener, one of the most sophisticated general staff officers of World War I, did a coherent defense policy run in accordance with resources available and the Republic's foreign policy.[7] But Groener's short tenure and the troubles plaguing the Weimar Republic in the early 1930s negated any lasting impressions that he might have made on either military strategy or organizational conceptions.

Adolf Hitler's first defense minister (after 1935 renamed the war minister), General Werner von Blomberg, attempted to establish a coherent military structure to support the rapid buildup of German forces after 30 January 1933. But Hitler had no intention of allowing such a concentration of power in the hands of one individual, while the three services willingly helped to thwart Blomberg's efforts. Moreover, given Blomberg's previous record and understanding of strategy, one may doubt his chances of successfully bringing strategic coherence to the Third Reich even with Hitler's wholehearted support.

After the Fritsch-Blomberg purge in January 1938, the Führer assumed the position of war minister himself, created a high command staff (the OKW, *Oberkommando der Wehrmacht*) with only coordination powers, and drew complete control of the larger strategic and military problems confronting the Third Reich entirely within his own province. Thereafter, Hitler added to the administrative chaos by eventually appointing himself commander in chief of the army (December, 1941), and even army group commander (September, 1942). By 1942, two entirely separate army staffs, the OKH, (*Oberkommando des Heeres*, the former army high command) for the east and the OKW for the Mediterranean and the west ran the ground wars, while the air force and navy ran their own separate wars. Only Hitler had a relatively accurate picture of the general situation—exactly the situation that he desired.

Many historians have blamed this organizational nightmare on Hitler, and given his inclinations, he surely deserves substantial credit for the administrative chaos. Nevertheless, from the onset of German rearmament, the three services balked at all of Blomberg's attempts to coordinate their activities. On one side, the army, jealous of its traditional role as the Reich's premier service, refused to tolerate interference by the defense ministry (and later by the OKW) in what it regarded as its private preserve, the formulation and conduct of the Reich's ground defense. On the other, Hermann Göring's unique position, not only as commander in chief of the Luftwaffe but also as the number-two man in the political hierarchy, insured that Blomberg exercised even less influence over that service. The navy sailed in the lee, protected by Admiral Erich Raeder's connection with Hitler's and the navy's own sense of independence.

The combination of these factors resulted in the German military being unable to wage war in a unified, joint fashion. Each service went its own way, contributing further to the failure of the German military to bring forward sound strategic advice.

In the Anglo-Saxon world a substantial portion of political effectiveness rests on the ability of military organizations to articulate their financial and resource needs. For the most part, the German military lacked such concerns. In fact, through 1911 the Prussian war ministry waged a successful campaign to limit the army's size despite the threatening international situation. Its stand reflected fears that a larger army would need a larger officer corps, one whose size would require more middle-class officers. Such a change, it feared, would seriously damage the position of the nobility within the Wilhelmine state.[8] Only in 1912 did the general staff and the war ministry reach a satisfactory compromise that provided sufficient troops to support the Schlieffen Plan.

Even during the Weimar Republic, a period that the German military regarded as hostile to their own interests, the problem was not that the state was unwilling to provide the resources that the services demanded. Rather, the limitations on military spending reflected the demands of the Treaty of Versailles; indeed the Republic and its ministers generally supported the military's efforts to evade the treaty's restrictive clauses. The extensive financial support, all of it secret, that the Republic provided to the army's clandestine ventures in the Soviet Union is a case in point.

With Hitler's coming to power, any previous limitations on military spending ceased. The Führer provided the three services (the Luftwaffe now appeared at the table) with a blank check that allowed them to embark on a massive program of rearmament.[9] The wisdom of a rearmament program with virtually no strategic cooperation among the services and within an economic framework desperately short of raw materials and foreign exchange, however, is open to question.[10]

Perhaps only within the German navy before the First World War did a German military organization establish an effective internal political coalition and manipulate the political situation to gain the resources needed to support its military designs. Grand Admiral Alfred von Tirpitz persuaded the Kaiser, the Junkers, and the industrialists to unite to support a radical buildup of the German fleet.[11] Only as Germany faced war in 1914 did the navy finally take a back seat to the renewed emphasis on the army. But by that time its buildup had made the High Sea Fleet one of the world's great navies. Its strategic significance and consequences for the Reich's continental position, however, were disastrous.

Germany's military institutions not only confronted few difficulties in acquiring financial resources, they also had access to the manpower and technological resources that modern military institutions require. Ironically, the prewar Prussian war ministry firmly opposed the army's expansion up to 1911. The limits placed on military manpower reflected a conservative desire to restrict the number of middle class officers and the conscription of city dwellers. But once the war began, these inhibitions ended, and the war ministry even allowed Jews to become officers. Since the social prestige of officership in the German army was considerable, the army attracted an excellent cadre to its officer corps.

German science and technology remained at the leading edge throughout the period. Surprisingly, with the availability of new technologies came a general unwillingness of German military institutions to use these advances effectively. Before World War I, the services largely ignored the technological advances taking place throughout European society. General Helmuth von Moltke, then chief of the general staff, commented in 1910 that the possibility of using aircraft to drop bombs was, "for the present unimportant."[12] While conventional munitions remained unsurpassed, what was difficult for the Germans to see, however, was the direction toward which technology was driving warfare. In World War I the Germans responded so late to the tank, introduced by the British on the Somme in 1916, that they never produced an effective model. In the next war, only after the Soviet T-34 had appeared on the eastern front did they respond with their Tiger and Panther designs; even then, failure to decide on technically feasible programs led to innumerable production runs and models.

In many respects German engineering and technological potential far exceeded the capabilities of their opponents.[13] But the failure to establish coherent systems to evaluate technological advances or to set priorities between competing systems often left the German military in the unenviable position of being equipped with inferior weapons (particularly in the air and sea war), while desperately waiting for something better.

Military organizations, at least in the western world, are supposed to be the servants of the state. One of the most disturbing aspects of the German military throughout this period concerns their penchant for mixing politics with an often astonishing naïveté. Before World War I, the Prussian military felt that the rising strength of the Social Democrats among the working class threatened their position and that of the monarchy. Kaiser Wilhelm II was given to bloodthirsty pronouncements about how he would settle accounts with this left wing threat. The military egged their Kaiser on, though he stopped short of waging war on the working class.[14] Nevertheless, military contempt for even the Bismarckian constitution led to the disastrous 1913 affair at Zabern in Alsace Lorraine, where the local garrison commander literally took the law into his own hands.[15]

World War I only reinforced the military's contempt for the civilian world. The *Frontsoldaten* believed that the world back home held little understanding of their sacrifices. Almost immediately after the crackup of November 1918 the military came to blame civilian society for the Second Reich's collapse.[16] Ironically, from the beginning of the war, the German military had guided policy and strategy. By 1916 General Erich Ludendorff and Field Marshal Paul von Hindenburg had established a dictatorship that excluded politicians and civilians from the decision making processes.[17] In the realms of neither strategy nor politics were these generals competent, but they did provide the German nation with a military ideal that the Kaiser no longer could fulfill. But while Ludendorff's capabilities in responding to the tactical realities of the battlefield were extraordinary, he was incapable of addressing the strategic and political problems of the war.[18]

Imperial Germany's collapse in 1918 and the military reaction to the ensuing chaos established the most dangerous of postwar military attitudes: a search for a new leader who could replace the Kaiser and provide the bonds of loyalty that the monarchy had failed to maintain during the war. This search would lead many in the officer corps to a wholehearted acceptance of Adolf Hitler's Führer state with its *Volksgemeinshaft*. For the time being, the German military had to live in a Republic with few of the beloved symbols of Imperial Germany. An uneasy alliance between the Weimar Republic's Social Democratic leadership and the old officer corps enabled the new Republic to eliminate the danger from the far left, but it also carried over the old regime's military institutions into the new "democratic" era. The Republic was the loser. Its defenders, now called the *Reichswehr*, carried on most of the traditions and prejudices of the old military.

The *Reichswehr*'s architect, General Hans von Seeckt, saw to it that the technocrats of the general staff rather than the nobility or the *Frontsoldaten* controlled the new officer corps. That decision insured that the German military would establish a doctrine and approach to combat on the tactical and operational lessons of 1918; unfortunately, it did not provide for any analysis as to where the strategic conduct of the war had gone wrong. Seeckt, generally unfriendly to the new regime, responded to Hitler's Munich Beer Hall Putsch by refusing to commit the *Reichswehr* to defend the Republic.[19] As he expressed it to his senior officers, the *Reichswehr's* loyalty lay with the "German state" rather than with the Republic. After the Third Reich's collapse in 1945, Seeckt's supporters claimed that he had sought to create a military above political concerns and that the resulting military institutions could not deal effectively with Hitler's Nazi regime. On the evidence this judgment is fundamentally flawed; in fact, Seeckt tried to distance the army from the Republic. Political involvement, at least in terms of vocal support for a more nationalistic form of government, certainly mirrored his conceptions.

Seeckt would not, however, have approved of the machinations of Kurt von Schleicher, who within the defense ministry played a crucial role in bringing Adolf Hitler to power. Schleicher almost singlehandedly forced the resignations of Groener, Chancellor Heinrich Brünning, and his successor Franz von Papen. Schleicher eventually manipulated Hindenburg into naming him as chancellor, but in so doing lost his standing with the president. It is hard to depict a more disastrous intervention in politics by the military than that of Schleicher.[20]

Ironically, in Hitler the German military found the leader that they had craved in the First World War—a man who could unite the German *Volk* behind the military and provide a rationale and ideology for German nationalism.[21] The first years of Nazi rule resulted in a honeymoon between the Third Reich and its military institutions. The government provided unlimited funding beyond the military's wildest expectations; and if Hitler forced the pace of rearmament faster than what seemed prudent to some, most happily pressed on with the tasks of the buildup. The regime did take extraordinary diplomatic and strategic risks in these early years: withdrawal from the League of Nations in 1933, declaration of rearmament in 1935, and occupation of the Rhineland in 1936. In each case, the lack of response by the other European powers confirmed Hitler's strategic assessments. The real break within the military came in 1938. In that year Hitler, taking enormous risks, pushed Germany to the brink of war with Czechoslovakia and the other major powers.[22] The resulting crisis divided the German military into two camps: the smaller represented by the chief of the general staff, Ludwig Beck, the larger consisting of the great majority of the officer corps. Beck argued that the risks attendant with Hitler's policy represented a political and strategic danger that the generals in their responsibility to the nation could not accept. But most generals thought that the political questions must be decided by the regime, a surprising change of heart compared to the attitudes expressed so often during the Wilhelmine Reich and the Weimar Republic. General Erich von Manstein, Beck's protégé, wrote his former chief in July urging him not to resign over the Czech crisis because "Hitler has so far always estimated the political situation correctly."[23] There was also a certain naïveté on the part of the chief of staff. Beck proposed to the generals in mid-July 1938 that they combine with the better, more responsible party leaders to eliminate the SS and party radicals, to restore justice, and to establish Prussian standards in government while leaving Hitler as head of state.[24]

After Hitler's stunning success at Munich, the military never again confronted the Führer over political or strategic issues. Rather they accepted, sometimes grudgingly, mostly enthusiastically, the Führer's political assumptions. In fact, Germany's disastrous approach to the problem of conquering the Soviet Union resulted from the generals' ready acceptance of the ideological framework within which Hitler cast "Case Barbarossa" (code name for the invasion of the Soviet Union).[25] The acceptance of Hitler's ideological crusade insured that the Soviet peoples would rally around Stalin's brutal regime and that the German army would cooperate wholeheartedly with the murderous activities of the *Einsatzgruppen*. In proclamations to the troops, several senior generals made clear what kind of war they were waging—one fully in accordance with Hitler's precepts.[26]

The overall impression left by the German military's participation in national politics is one of naïveté. Their negative contributions added immeasurably to the national disasters of 1918 and 1945.

Strategic Effectiveness

According to Clausewitz's much quoted aphorism, "When whole communities go to war—whole peoples, and especially civilized peoples—the reason always lies in some political situation, and the occasion is always due to some political object . . . . It is clear, consequently, that war is not a mere act of policy but a true political instrument, a continuation of political activity by other means."[27] Unfortunately, the German military consistently failed to recognize that strategy demands an intelligent relationship between political ends and available means. Strategy as cast by the German military almost always reverted to operations; rather than allowing strategic and policy concerns to guide operations, military operations became an end in themselves. This constant confusion of strategy with operations helps to explain the German military's approach to the First World War and their incapacity to address the flawed means to ends relationship in Hitler's conduct of World War II. The Schlieffen Plan is only one of the more glaring examples of this weakness. Confronted by complex diplomatic, strategic, and political problems, namely Germany's vulnerabilities on the European Continent with the threat of a two-front war, Schlieffen reduced Germany's options to a simple, massive plan that aimed to knock the French out with one blow and then to solve the Russian problem at leisure.[28] Schlieffen saw little need to coordinate his plan with the German foreign office and had little desire to contemplate the political and diplomatic consequences of violating Belgian neutrality, to consider the impact of a British intervention, or even to inform the Prussian war ministry responsible for supplying the troops of his plans. Thus, the general staff reduced political, diplomatic, strategic, and even logistic problems to a simple operational formula—one that resulted in a European disaster.[29]

Part of the problem lay in the fact that the German military had by the end of the nineteenth century reached the conclusion that Clausewitz was passé. Throughout the twentieth century, they regarded Clausewitz's writings as irrelevant. As General Leo Geyer von Schweppenburg noted after World War II on general staff "education" before 1914: "You will be horrified to hear that I have never read Clausewitz or Delbrück or Haushofer. The opinion on Clausewitz in our general staff was that of a theoretician to be read by professors."[30] There is no evidence to suggest that the interwar officer corps had any different perceptions or interest in what Clausewitz had to say.[31]

This confusion of strategy with operations is suggested even more vividly by Ludendorff's reply to a question by Crown Prince Rupprecht of Bavaria who wanted to know what the objective of the March 1918 "Michael" offensive might be. Ludendorff replied: "We will punch a hole into [their lines]. For the rest, we shall see. We also did it this way in Russia!"[32] Without a strategic or political vision the Germans operated in a vacuum, one in which they established military goals with regard

only to military criteria, irrespective of strategic or political consequences.

The measure of a military organization's strategic effectiveness lies
in its capacity to calculate acceptable risks over the long-term. Because
the German military so consistently opted for short-term operational
gains, the long-term consequences were often catastrophic. With its
lack of concern over British participation in a European war, the Schlieffen Plan is an excellent example, but the German navy used equally
flawed reasoning in urging Hitler to declare war on the United States
in summer 1941.[33] Here, the *Kriegsmarine* argued the issue of war on the
basis that a declaration would allow its U-boats to ravage the eastern
coastal waters of the United States. As for the long-term economic consequences of such a decision, the navy does not seem to have had a
concern.

Even more seriously, the Germans tended to evaluate avenues of
approach without fully considering whether the benefits of gaining
strategic objectives outweighed the costs of failure. The exception was,
of course, Beck's objection to Hitler's inordinately risky policies toward
Czechoslovakia in summer 1938. Beck had no doubts that the Czech
Republic represented an irritant that Germany needed to be rid of, but
the consequences of a 1938 invasion of Czechoslovakia, he argued,
would result in Germany once again finding herself ranged against the
rest of Europe; therefore Germany must not take the plunge, the risks
being simply too great.[34] But few other generals had any interest in
supporting the chief of staff. More typical was the approach to the
Schlieffen Plan where the risks and consequences of failure were simply
ignored. Schlieffen thought that the consequences of failure posed concerns only for politicians. Ironically, Hitler mirrored this attitude; as he
once argued, it was his business as national leader to calculate the risk
and the business of the generals to fight the resulting wars. Consequently, where the military had floundered in confusion in World
War I, preventing civilian authority from assessing risks and placing the
operational cart before the strategic horse, they now turned the assessing of risk over to a political leader whose strategic equation substituted
personal will whenever the means to ends equation seemed out of balance.

This German approach to war from the bottom up resulted in an
inability to connect strategic goals with available forces. Groener con-
veyed some sense of this in a critique of the strategic conduct of World

War I that he delivered to his officers in early 1919. He suggested that the Reich's defeat had resulted from its attempt to become a world power before it had secured its position on the Continent. By attempting both at the same time, Germany made her defeat inevitable.[35] But Groener's realism, a mark of his tenure as defense minister, did not represent the view of most officers even in the 1920s when the Republic possessed minuscule military power. More typical was a war game designed by the *Reichswehr* in late 1922 that posited a combined effort with the navy (now reduced to a predreadnought force) to thwart a Franco-Danish invasion of northern Germany. The French soon underlined the lack of realism in this game by occupying the Ruhr—an action to which the Reichswehr had no possible reply except national suicide.[36] And one must not forget the almost universal belief within the officer corps that the army had stood unbeaten in the field in fall 1918, only to be stabbed in the back by Jews and socialists, the so-called *Dolchstoss* legend.

Among other things, strategy demands consideration of alternatives; if one's assumptions prove faulty, then one must adapt to the real conditions. Neither army, navy, nor Luftwaffe had this flexibility. The navy proved the most dogged in its refusal to adapt its strategic assumptions to reality. As Holger Herwig has suggested:

> Seapower, in a word, consists of fleet and position: one is useless without the other. Tirpitz either ignored or never grasped Alfred Thayer Mahan's unwritten presupposition that unfettered access to the world's ocean was the cardinal prerequisite for seapower. Given that Britain was Germany's primary potential opponent [especially in view of everything that Tirpitz was striving to achieve], a brief glance at the map should have confirmed the obvious: the British could bottle up the German fleet, based either in Kiel or Wilhelmshaven, in the North Sea, if they chose to close the Straits of Dover and the waters between Scotland and Norway. Despite this, Tirpitz failed to develop an alternative strategy.[37]

But Tirpitz's greatest mistake came on the political-strategic level; his "risk theory" rested on three flawed assumptions: that in the long run the British would not match either the manpower or the financial costs of a naval race; that the British would remain isolated and could never come to terms with the French and Russians; and that Germany could afford the costs of both continental defense and a naval buildup. By 1907, those assumptions looked increasingly dubious. The British showed their willingness and ability to maintain naval superiority; the *Dreadnought* class battleships placed the cost of the naval race beyond the Reich's capacity to support; and finally, the ententes with France (1905) and Russia (1907) placed Britain squarely within a great anti-German coalition. Did any of this serve to change Tirpitz's strategic assumptions? Certainly not on the basis of continued German naval policy.[38] Tirpitz's question to the fleet chief in May 1914—"What will you do if they do not come out?"—underlines the bankruptcy of German naval strategic policy.[39]

But the German army also proved unwilling to prepare for contingencies or to adapt strategy to realities. In fact, the strategic approach that resulted in the Schlieffen Plan aimed at removing contingencies and leaving only one avenue open for German strategy should a crisis arise. The army also would not adapt its assumptions to the real strategic conditions of war in its 1941 campaign against the Soviet Union. German planning for Barbarossa rested on several assumptions, few of which, even given the stunning nature of German victories early in the campaign, proved close to the strategic mark. The initial onslaught did not destroy the Red Army in the frontier areas; the Soviets put far more troops into the field than the Germans expected;[40] and Stalin's regime, helped by the murderous crusade that the Germans waged, weathered the storm created by the invasion. Nevertheless, despite the lateness of the season and serious threat of oncoming winter, the German high command, Field Marshal Walter von Brauchitsch, General Franz Halder, Field Marshal Fedor von Bock, and their staffs persisted in a maniacal drive in October and November to capture Moscow. They refused to address any alternatives because of the pressure of assumptions that they had made at the beginning.[41]

The achievement of strategic goals by operational means also demands some connection with logistical capabilities and the industrial support that the national economic structure can provide. In terms of war in the industrial age, this may be one of the most important factors in the means-ends equation. Yet it is here the Germans appear weakest in their approach to war both strategically and operationally. The Schlieffen Plan rested on the logistic assumption that the Belgian railroads would be available. They were not, and the German supply system could not support the drive deep into France—the railheads remained far back in Belgium as German infantry staggered up to the Marne.[42] But it was during the 1941 Russian campaign that German logistic failings had their greatest impact on the strategic outcome. In the planning stages of Barbarossa in fall 1940, General Friedrick Paulus, future commander at Stalingrad, made clear that the Wehrmacht might confront a logistic nightmare in the depths of Russia. Logistic calculations indicated that German supply services would have difficulty provisioning invading forces much beyond a depth of 500 kilometers east of the frontier—a distance well short of Leningrad, Moscow, and the Donets Basin.[43] But such a significant warning did not alarm the senior leadership; even though railroad repair and conversion represented the essential support element in the campaign, those units charged with that task remained at the bottom of the army's priority list in the movement forward.[44]

Another significant German deficiency in both world wars lay in its approach to industrial mobilization. Performance in World War I was spotty at best; the so-called Hindenburg program that Ludendorff cast in late 1916 was a general economic disaster because it set such unrealistic goals and was so far removed from the realities of available capacity.[45] But the real weakness in the military's approach to production and its relation to strategy showed even more clearly in the next war. Admittedly, some officers recognized that industrial preparation for a long war would play a crucial role in a future conflict. But in the late 1930s the Germans could not prepare the economy for a long war and at the same time meet Hitler's inordinate demands for the buildup of the Wehrmacht. Through the French campaign of 1940, economic weaknesses placed severe constraints on what industry could produce for the rapidly expanding Wehrmacht,[46] though little connection existed between the Führer's strategy, the Wehrmacht's preparations for war, and the degree of economic preparation and mobilization.[47] German rearmament, massive throughout the late 1930s, caused such severe economic strains by 1938 that it became a major factor in driving Hitler to take increasingly dangerous risks.

But the conquest of France in May-June 1940 along with the booty of Scandinavia and the Low Countries put the Third Reich in a secure economic position. It could now mobilize the entire European Continent; luckily, it did not. Instead, while Britain, the United States, and the Soviet Union made desperate efforts to increase production, the Germans remained confident and unhurried.[48] Admittedly, throughout the war, Hitler's unwillingness to pressure the German people for fear of a replay of the 1918 revolution represented a major stumbling block to fuller economic mobilization. But much of Albert Speer's success rested on the fact that the German war economy drew on the resources of the rest of Europe. Had the Germans mobilized Europe's latent economic strength after the defeat of France, the problems of their opponents would have been magnified. However, they did not do so, and while most of the blame rests on Hitler, the military shares in the failure.

The final test of strategic effectiveness involves the capacity of military organizations to work with allies and to utilize allied strengths to the coalition's advantage. Again the Germans did an extraordinarily bad job. In World War I, Germany's chief ally had been the Austro-Hungarian Empire. Before 1914, the Germans failed to notify the Austrians of the ramifications of the Schlieffen Plan; moreover, there was virtually no coordination between the two powers in the prewar period. Only at the outbreak of the war did the German military attaché in Vienna cable Berlin to suggest that "it is high time that the two general staffs consult now with absolute frankness with respect to mobilization, jump-off time, areas of assembly, and precise troop strength."[49] Such a telegram suggests a complete lack of previous coordination; the course of World War I saw little improvement. In 1916 the Austrians weakened their forces on the eastern front to mount an offensive against the Italians; that in turn almost resulted in a complete collapse of the eastern front when the Russians attacked the weakened Austrian positions. To save the situation in the east the German high command had to shut down the battle of Verdun.

In World War II, the Axis, a supposed alliance between Nazi Germany, Fascist Italy, and Imperial Japan, represented little more that an agreement between thieves whose common interest lay only in their desire to steal what belonged to others. The Axis powers did no common planning before the war and precious little during it. Hitler made his decisions to sign the Nazi-Soviet Non-Aggression Pact and to attack Poland without consultation with his allies. The Italians fought their own "parallel war" in the Mediterranean until military catastrophes forced them into the tight embrace of the Germans.[50] Never true allies, the Germans and the Japanese made no coordinated plans, no effort to help each other, and had no sense of how they might work together to affect the course of the conflict for their mutual welfare. Undoubtedly, geographic distance contributed to the weakness of the bond, but one also suspects that the Germans had neither the interest nor inclination to think in terms of larger alliance issues. Given Nazi attitudes toward other nationalities and races, this is not surprising.

Operational and Tactical Effectiveness

Moving down a notch from strategy to German battlefield performance in a variety of theaters and conditions, one moves into a very different realm of military competence. But despite substantial differences between the services, their strengths and weaknesses remained surprisingly similar. Above all the Germans displayed a consistent capacity to adapt to the intricate difficulties of modern war, though they lacked any spirit of joint cooperation. Before 1914, the army and navy could not draw on a body of common historical experience. The operational plans of that period reveal little common ground or understanding: the army executed the Schlieffen Plan and wheeled inside of Paris, while making no effort to capture the Channel ports. In retrospect, a few cavalry divisions might have brought that crucial area under German control. On the other hand, the navy did not interfere with the transfer of the British Expeditionary Force to the Continent; those British troops played a significant role in arresting the German drive on the Marne in September and in protecting the Channel ports in fall 1914. Four years later when Ludendorff launched his "Michael" offensive on which Germany's bid for world power rested, the navy again did nothing; its battleships remained securely anchored in the North Sea harbors. Consequently, the British found themselves able to rush reinforcements and supplies across the Channel to their hard pressed forces in northern France.

Much of the failure to cooperate resulted from the lack of a coordinating high command. Moreover, the services harbored deep mutual suspicions toward each other, the navy even refusing to exchange intelligence data with the army before the outbreak of war.[51] With no tradition of cooperation and with fundamentally different *Weltanschauungs*, the two services found cooperation almost impossible during World

War I. Nevertheless, as Gallipoli underlined, joint matters went no more smoothly on the other side. But in the interwar years the British, unlike the Germans, created the chiefs of staff system that eventually made effective joint planning possible. As one of the leading officers in the OKW noted at the end of the war: "in fact the advice of the British Chiefs of Staff and the U.S. Joint Chiefs was a deciding factor in Allied strategy. At the comparable level in Germany there was nothing but a disastrous vacuum."[52]

In the next war, relations between navy and the new service, the Luftwaffe, got progressively worse. Göring lacked any interest in participating in cutting off Britain's sea lines of communications, nor was the Luftwaffe's general staff enthusiastic. Consequently, the air units participating in the maritime campaign represented a small percentage of the Luftwaffe's overall strength. The navy, for its part, often failed to coordinate air reconnaissance with the disposition of its U-boats. As a result, the navy's U-boat wolfpacks posed the sole threat to the Allied sea lines of communication for much of the war.[53]

The relationship between the army and navy can best be described as tenuous. In early June 1940, even before planning for "Operation Sea Lion" had begun, the navy sent out its last major fleet units, the *Gneisenau* and *Scharnhorst* off Norway's North Cape, not for operational or strategic purposes but rather to influence postwar budget debates.[54] The planning for "Sea Lion" suggests the level of cooperation between the services; the OKH drew up plans that entirely disregarded the navy's actual strength, with the navy being equally unrealistic in planning to land the army on a minuscule section of the British shores.[55] The Luftwaffe considered "Sea Lion" irrelevant, since it believed that its own aerial campaign would win the war against Britain and make such a landing unnecessary.[56] As for realism in the OKW, the supreme headquarters, General Alfred Jodl, operations officer, noted in his diary at the end of June that final victory over England was only a question of time.[57] The OKH initially regarded the problems involved with "Sea Lion" as no worse than those posed by a "large scale river crossing."[58]

But the German military could work in the joint environment when operations occurred at lower levels of command. "Weserübung" (code name for the invasion of Denmark and Norway) represented the best joint performance of German military forces in the war. The Luftwaffe dominated the seas around southern Norway, while the navy slipped the army into the major Norwegian ports. Where Norwegian defenses blocked naval forces in the Oslo fiord, Luftwaffe paratroopers seized the main airport outside the capital, and transports landed sufficient reinforcements for the invaders to seize Oslo by the early afternoon.[59] Similarly, the seizure of Crete by Luftwaffe paratroopers and army mountain troops indicates that the services could skillfully cooperate.

The one area where interservice cooperation reached levels that remained unmatched by Germany's opponents until well into the war had to do with the development of close air support doctrine by the army and Luftwaffe. Unlike most other air forces in the world, the Luftwaffe devoted considerable assets to ground forces support and to insuring that those air units received proper training and equipment.

At this point, the three services' effectiveness is best examined in their separate operational arenas, starting with the army, given its position within the German military hierarchy. Of all German military institutions, the army reached the highest levels of battlefield excellence and performance, yet entered World War I with serious deficiencies that almost brought defeat in the first two years. In 1914 the general staff exercised sway over only a small portion of the decisionmaking processes; in particular it drew up war plans and had some influence over tactical and operational questions. The Prussian war ministry exercised complete control over manpower questions as well as budgetary and administrative issues. Significantly, the general staff could not get Lieutenant Colonel Georg Brückmüller, the great artillery expert who designed a significant proportion of the bombardment plans in the last war years, promoted to the rank of permanent colonel.[60]

The course of the fighting through to 1916 did not suggest German battlefield superiority. The Schlieffen Plan, whatever its intellectual power, represented a significant departure from the operational art of Moltke's approach to the campaigns of 1866 and 1870. In the latter case, army commanders had great latitude to react to the real situation that confronted them rather than to follow obsolete orders. Consequently, Germany reaped the benefits of independent actions taken by commanders on the spot. However Schlieffen's tightly controlled plan left little room for deviation. In effect his plan represented an attempt to contradict Moltke's aphorism that "no plan survives contact with the enemy."[61] To a great extent the defeat on the Marne resulted from the failure of a plan that directly contradicted the operational art and ignored the conditions of ambiguity that govern the conduct of warfare.

The performance of German troops on the tactical level during this early period matched that of their opponents. The "slaughter of the innocents" in October 1914 threw ill-trained university students against the British and French in Flanders and represented as great a tactical disaster for the Germans as the French suffered in the opening battles along the frontier or the first day on the Somme for the British in 1916.[62] By 1916 the Germans faced a desperate situation. Their attack on Verdun had failed miserably. After a disastrous start on the Somme on 1 July 1916, the British army on the offensive imposed a one-to-one ratio in battlefield losses on German defenders who continued to pack frontline trenches with the bulk of their infantry, exposing them to British artillery fire.

The Kaiser then brought in Hindenburg and Ludendorff to rectify a desperate situation. On the strategic level, these two compounded Germany's problems by agreeing to naval demands for the unrestricted submarine warfare that brought the United States and its economic potential into the war. But for the first time in the war a senior military leader, namely Ludendorff, did address the considerable tactical problems that the war had raised. Several significant studies examine how the Germans addressed these difficulties.[63] For our purposes the crucial issues do not involve the solutions but rather the paths to those solutions and the consequent retraining of the army. First, Ludendorff used the general staff system to insure a reliable flow of information from the battlefield to the high command. Second, he himself went out to the western front and discovered that conditions differed significantly from those on the eastern front. His memoirs note that, "it was my duty to adapt myself to [those conditions]."[64] In his fact finding tour, Ludendorff demanded that the soldiers and staff officers who reported to him speak their minds, not pass along something "made to order."[65]

In effect the Germans invented modern defensive warfare and operational concepts; they created a defense in depth that would exhaust enemy attackers with successively stronger positions that placed infantry and artillery reserves beyond the range of enemy firepower. Now the defense could gauge the locations of significant enemy penetrations; it could then launch counterattacks and seal off enemy penetrations. In 1917 these new defensive tactics, encapsulated in the doctrinal manual, "The Conduct of the Defensive Battle," allowed the Germans to blunt the Nivelle offensive in the spring, a victory that came close to destroying the French army. Then, on the basis of lessons learned in these defensive battles with their supporting counterattacks, the Germans developed modern, decentralized, mission-oriented, offensive tactics and the operational concepts that allowed them to launch their devastatingly effective attacks in spring 1918.

Several factors account for these German successes. The key to modern battlefield performance is decentralized control. Units down to company and platoon level must be prepared to act on their own. Given the ambiguities of the modern battlefield, higher command levels can provide general guidance at best and then demand that commanders and troops on the scene act in accordance with that guidance. Such an approach depends on a high level of leadership at all levels; it demands that soldiers and commanders receive coherent, stringent training; and it demands a coherent doctrinal approach. Throughout this period the Germans achieved these standards through a variety of means. The general staff system allowed a relatively free flow of information up and down the chain of command. Consequently, the Germans adapted more quickly and more coherently to changes on the battlefield. The general staff system also allowed a more thorough integration of the combat branches because general staff officers acquired a thorough grounding and knowledge of them. Finally, the Germans took their combat doctrine seriously. Once the doctrinal experts in the general staff had evolved "The Conduct of the Defensive Battle" in late 1916 and the new offensive doctrine by late 1917, the army and the general staff established schools throughout the western front's rear areas to insure that commanders at all levels understood and complied with the new doctrine.[66]

But by the end of 1918, *all* the armies on the Western Front had adapted to the new conditions of offensive and defensive warfare. As the French defense in front of Rheims in July 1918 and the British attacks in the late summer showed, the Allies learned from the Germans how to utilize defense in depth and employ new offensive tactics. In fact, the British further improved the German system by including the tank in their attacks. However, when peace came, only the German army coherently and consistently carried over the lessons of the war into the interwar period. By 1930 the French had forgotten the new direction that tactics and doctrine had taken in 1918.[67] For their part the British failed to learn the lessons of the war until it was too late: not until 1932 did they establish a committee to study the lessons of the last war.[68]

The German army did not make this mistake. Historians often berate military institutions with the charge that "Colonel Blimps" consistently prepare for the last war. This charge misses the point that most armies fail to learn the significant lessons from the last war. In this case only the Germans prepared for the next war on the basis of a realistic appreciation of what had happened in the last one, precisely the reason why they did so well on the battlefield in World War II. By 1924 the disguised general staff (which the Treaty of Versailles had outlawed) had prepared a doctrinal manual, *Die Truppenführung*, that distilled the tactical lessons of the war into a clear and coherent form. On that excellent basis, the army prepared itself and the troops under its command for the remainder of the interwar period. This doctrine, based on a combined arms approach, emphasized the utilization of all combat branches to insure surprise, exploitation, decentralized planning and execution, and above all, speed. Artillery officers, infantrymen, and support branches could now think within the same intellectual framework; consequently, in the 1930s the German officer corps could realistically incorporate the new armored force. There was nothing revolutionary about the employment of armored units by the German army in 1940: all the doctrinal pieces already existed.

The coherence between German doctrine in 1918 and 1939 represented the triumph of the general staff after the end of World War I. In 1920 the Germans faced the demands of the victorious powers that they drastically reduce the size of their officer corps. The army commander in chief, Seeckt, brought virtually the entire general staff over from the wartime army with a sprinkling of combat heroes and regular officers to fill out the force. Most officers in the latter two categories found themselves separated from the army.[69] Moreover, Seeckt insured that the general staff became the dominant force in the new army.[70]

What characterized German ground forces in the 1930s was the army high command's recognition of the major flaws and weaknesses within its rapidly expanding organization. From the Anschluss through to the French campaign, it never used operational success as a standard of measure. No matter how stunning the Polish campaign might have appeared to outsiders, the OKH felt most dissatisfied with the army's performance in the campaign and introduced a drastic, thorough retraining program to correct those deficiencies.[71] Moreover, the German army actually learned from its experiences on the battlefield.[72] To evaluate combat potential, the OKH, army group, army, corps, and division commanders expected subordinate commanders to give honest, accurate reports on troops' capabilities. If a unit appeared substandard, its commander must say so and take steps to correct its deficiencies. The higher the level of command the more critical and demanding the expectations.[73] Besides its realism and toughness, German military training also gave the German soldier a high sense of unit identification and a belief that he could depend on the soldier next to him. Finally, the OKH and the general staff imposed their high standards and expectations throughout the officer corps and enlisted ranks.

The Germans fused this effective reporting system into a coherent, realistic doctrinal scheme. In 1939 they pushed the armored forces to maximum operational capabilities precisely because of their coherent combined arms doctrine.[74] The aftermath of the Polish campaign shows the many connections between thorough analysis, realistic doctrinal adaptation, and ruthless training in accord with the new doctrine.[75] After the conquest of Poland, the OKH demanded that army units down to regimental level render accurate "after action" reports. Within a month those reports had been gathered, digested, and established as the basis for a complete retraining of the army. This retraining gave doctrinal adaptation its fearsome edge. Tireless preparation, lasting from November 1939 to April 1940, turned the German army into an instrument of power that wrecked the western front in a matter of days.

But the historian of German military effectiveness would not be accurate in suggesting that German battlefield performance rested on purely military factors, particularly in World War II. In that conflict, ideology substantially motivated troop performance, which the army leadership understood even before the war. As early as 1938 the army's commander in chief, Brauchitsch, specified that "the officer corps must not allow itself to be surpassed by anyone in the purity and the genuineness of its National Socialist *Weltanschauung*."[76] Two years later he wanted "not the slightest doubt about the fact that the training of the soldier to be a determined and aggressive fighter [can] not be separated from a lively education in the National Socialist sense."[77] The performance of German ground forces on the eastern front, moreover, intimately linked ideological preparation and motivation with combat readiness.[78] The very factor that played an important role in the strategic ineffectiveness of the invasion of the Soviet Union, namely Hitler's ideological crusade, also helps explain why the German soldier fought so well.

But the army's performance showed serious weaknesses in its support structure, particularly supply and intelligence, crucial factors in any modern definition of military effectiveness. In addition to the many crucial logistical mistakes on the strategic level that so bedeviled the German conduct of war, the logistical mistakes and their consequences had an equally serious impact on the operational level. The Schlieffen Plan failed as much from its logistic weaknesses as from any other reasons.[79] But failures in logistic planning and execution had the greatest consequences in World War II. Only desperate expedients allowed the Germans to equip the forces for "Barbarossa" with their basic TO&E (Table of Organization and Equipment) levels. To do this the Germans had to give even elite panzer and motorized infantry divisions a mixed bag of weapons and support vehicles, many of which had been captured from the British and French. More than half of the invading divisions either had shortages of equipment or used equipment captured during the western campaigns of 1940.[80] French, British, Belgian, and Dutch military and civilian trucks jostled with Norwegian mountain artillery and Czech tanks to create a logistician's nightmare. Among other bizarre supply estimates, the Germans calculated that they would not need more ammunition than they had used in the invasion of France, surely one of the greatest mis-calculations of military history.[81] Even units charged with repair of the main Brest Litovsk-Smolensk rail link, essential for the buildup of supplies required for resumption of an advance on Moscow, received the lowest priority in the move forward.[82]

But of all the logistic miscues of 1941, the advance on Moscow surely represented a mistake with the most disastrous operational consequences. By late September 1941, the supply system had finally reached a point where an advance could resume. However, it could only do so by sacrificing the buildup of supply dumps of ammunition, fuel, food, and winter clothing that German troops would require for the Russian winter. The OKH and Army Group Center opted for a resumption of the offensive even though they fully understood the possible supply consequences. Moreover, they do not seem to have consulted Hitler; the disastrous defeat in front of Moscow was a direct result.[83]

The second area of support where the Germans showed themselves markedly inferior to their opponents lay in the field of intelligence. Like the other three services, the army's radio traffic, encoded on the enigma machine, proved highly vulnerable to the code breaking efforts of British intelligence at Bletchly Park.[84] In the early part of the war when much of the army's communications went by land lines and when German armies were rampaging through Europe, this posed no significant compromise. By 1942, however, the results were more damaging; information provided by these Ultra decrypts played havoc with movement of supplies to North Africa. Ironically, Ultra provided the only clear source to the western allies as to what was transpiring on the eastern front. By the time of the Normandy invasion, Ultra provided Allied commanders with intelligence on the location of headquarters of Panzer Group West (taken out soon thereafter by Allied fighter bombers) and the timing, direction, and purposes of the Mortain counterattack.[85]

But the most disastrous German intelligence failures came on the eastern front. There, at least from 1942, Soviet *maskirovka* (deception) misled German intelligence as to the weight, direction, and purpose of every major offensive that the Red Army launched as well as diverted attention toward attacks that never materialized.[86] For all of the vaunted fighting power of German troops on the Eastern Front, Soviet deception negated every battlefield performance advantage that the Germans might have enjoyed.

Like the army, the Luftwaffe enjoyed a high standard of performance on the operational and tactical levels. Of all the air forces fighting World War II only the Luftwaffe possessed a broadbased doctrine that took air power beyond the confines of "strategic" bombing.[87] It was not that the Luftwaffe did not take "strategic" bombing seriously. It did; only the Germans had prepared seriously to meet the navigational and technological demands of such an air campaign.[88] But the Germans had also learned from their experiences in the Spanish Civil War that hitting targets accurately from the air and the staying power of civilian populations under aerial assault presented difficulties glibly passed over by "strategic" bombing theorists. Luftwaffe air doctrine recognized the far wider contributions air power could make to reconnaissance, interdiction, air superiority, close air support, air defense, and air transport.[89]

In general, the Luftwaffe prepared itself better than other air forces to contribute to undertake joint operations. Its contribution to the advance and victories of German ground forces was an obvious mark of effectiveness in the early triumphant years. It proved a tenacious opponent on the wide frontiers that the Wehrmacht defended thereafter; and in the great day and night battles of 1943 and 1944 against the Combined Bomber Offensive, despite its being heavily outnumbered, the Luftwaffe came close to victory.

But severe deficiencies weakened the Luftwaffe's structure. On the operational plane, the Luftwaffe high command did not focus on clearly defined objectives in the Battle of Britain; consequently it moved from one target set to another. The high command remained unrealistically optimistic throughout the war. Admittedly, a substantial part of the problem lay in Hermann Göring. The *Reichsmarschall*'s general incompetence and his toadyism toward the Führer, not to mention his corruption, provided a continuous series of disastrous leadership decisions. But if Göring represented much of the problem, others also shared the blame. The young chief of the Luftwaffe's general staff, *Generaloberst* Hans Jeschonneck, fell under Hitler's spell from the beginning. In 1939 when confronted by Hitler's megalomaniacal demand that the Luftwaffe expand fivefold—hardly a realistic goal since such a force would need 85 percent of the world's production of aviation gas—Jeschonneck commented to the staff that "in my view it is our duty to support the Führer and not work against him."[90] In 1943 Johannes Steinhoff reported to Field Marshal Albert Kesselring on his way to command a fighter wing in Tunisia and to confront a disastrously deteriorating situation. He recounts the atmosphere in the field marshal's headquarters in the following terms:

> Never in my life will I forget the Air Fleet situation conference which I was permitted to attend. There was I, a combat officer, witnessing the prognostications and synthetic portrayal of the future course of the battle of North Africa. . . . Was I allergic to the high command and to staff officers? . . . I found their foppish affectation and general superciliousness insufferable. In their presence anyone honored by an interview with the field marshal would be made to feel like a clumsy carthorse; . . . When it was over we sat down at a large table to an excellent dinner presided over by the field marshal himself. "As soon as you've familiarized yourself with the theater," he began, "it is essential that you convince the people in your group that North Africa must be held—held at all costs. We shall be reinforcing the bridgehead and narrowing the front so that the position can be held without difficulty."[91]

With such attitudes it is not surprising that Kesselring would encourage his aircrews to show Japanese fanaticism to overcome Allied superiority in numbers.[92]

But it was the support area that most hurt the Luftwaffe. On the strategic level its leadership, with the exception of Field Marshal Erhard Milch, dismissed the threat represented by America's production capacity. Even as late as March 1942, Jeschonneck objected to a Milch proposal that the Luftwaffe raise production of fighters to 360 per month with the comment that he would not know what to do with so many aircraft.[93] By the time that the leadership awoke to the threat, the roof was falling in on the Reich.

On operational and tactical levels, the weaknesses in support areas appear just as glaring. As the leading historian of the Luftwaffe suggests, the Luftwaffe remained so focused on operations that engineering, logistics, and other support areas never received adequate attention.[94] At best its supply system could support a war confined to central and western Europe. Once the war expanded into the depths of the Soviet Union and from the North Cape to North Africa, the Luftwaffe found itself in desperate shape. By December 1941 the operationally ready rate for bombers had fallen to 32 percent and for fighters to 52 percent.[95] In traveling around Russia in fall 1941, Milch found aircraft needlessly abandoned because of breakdowns in the supply system.[96] While Milch fixed some of the deficiencies in the system, logistics remained a field that the high command took for granted, a fact which had a disastrous impact on the Luftwaffe's ability to fight a sustained battle.

Like the army, the Luftwaffe shorted its intelligence services in favor of operational capabilities. The undervaluing of intelligence and underestimation of enemy capabilities marked Luftwaffe operations throughout the war.[97] The basic intelligence study that guided the Luftwaffe's conduct of the Battle of Britain (produced by Luftwaffe intelligence on 16 July 1940) managed to get wrong virtually every factor of importance in the coming battle.[98] It estimated Hurricanes and Spitfires as being inferior to the Bf 110, made no mention of the RAF's radar defenses, and ended on the optimistic note that "the Luftwaffe, unlike the RAF, will be in a position in every respect to achieve a decisive effect this year." Not surprisingly the initial Luftwaffe report for the coming campaign estimated that it would take only four days to defeat Fighter Command in southern England, followed by a four-week period during which its bombers and long-range fighters could mop up the remainder

of the RAF and destroy the British aircraft industry.[99] Intelligence estimates on the strength of the Red air force before the beginning of Barbarossa were as unrealistic as the army's estimates on the Red Army;
Jeschonneck shortly before the invasion exclaimed: "At last, a proper
war!"[100]

The impact of Ultra on Luftwaffe operations ultimately spelled disaster. From its ability to suggest the weaknesses in the German
economic system to its warning that German fighter units were arriving
on airfields in France shortly after the beginning of "Overlord," Ultra
gave the Allied air forces an unparalleled view into the conduct of German air operations.

The navy's effectiveness throughout both world wars remained consistent. On the operational and tactical levels, the conduct of surface
operations in World War I was unremarkable; for most of the war the
High Sea Fleet remained in harbor contributing little to the defense of
the Reich. When the fleet did come out, such as at Dogger Bank in 1915
or at Jutland in 1916, its performance was lackluster; the most positive
comments one could make about the navy was that its ships could
absorb a great deal of punishment. At Dogger Bank, German battle
cruisers found themselves badly knocked about: Admiral Lord Beatty's
force sank the *Blücher*; the *Seydlitz* came close to blowing up, but she
flooded her magazines at the last moment, which allowed the Germans
to discover the danger of flash explosions in the gangways leading from
the magazines to the turrets. That discovery saved the High Sea Fleet a
number of ships at Jutland though two of the German battlecruisers
returned to harbor with their front decks awash. The real significance
of Jutland lay in the Royal Navy's continuing dominance over all entr-
ances and exits to the North Sea. The High Sea Fleet remained in port
for virtually the remainder of the war.

The real threat to the Allied strategic position was the undersea
attack on Great Britain's sea lines of communications; the Germans
launched their first assault in 1915 with completely inadequate means
(only twenty-eight boats of which half were ineffective gasoline pow-
ered U-boats).[101] When the Germans resumed unrestricted submarine
warfare in early 1917, they still only possessed 100 U-boats, which
meant only about thirty-five could stay on station at any one time. Over
the course of the next year they added only ten boats to their total
(eighty-seven constructed, seventy-eight sunk). Nevertheless, the naval
high command remained hostile to the new force despite their advocacy

of unrestricted submarine warfare. Admiral Eduard Capelle even suggested that an emphasis on U-boat construction would endanger the capital ship program.[102] Clearly he and his fellow admirals worried about what would happen to their role in a navy dominated by submarines.

In the interwar period, the German navy dreamed of recreating the High Sea Fleet and redeeming the shame of the revolution that had broken out on its ships and the scuttling of the battle fleet at Scapa Flow. Like the Royal Navy, the Germans studied for another Jutland; when Hitler provided large funding increases in 1933, the navy again opted for a big ship, big gun fleet. It showed little interest in aircraft carriers, which Raeder described as "only gasoline tankers."[103] As for submarines, the senior leadership believed the obituaries on that weapon that appeared in British publications. A 1939 study on the naval situation reported that "the importance of U-boats has considerably declined compared to 1915. One can assume that England has good detection gear which makes torpedo attacks on a secured unit or on a convoy impossible."[104] When war came in September 1939, the German surface fleet was still minuscule in size (three pocket battleships and two battle cruisers) and the U-boat fleet numbered only twenty-eight ocean-going submarines.

Condemned to fight on uneven terms, the Germans waged a tenacious struggle that came close to breaking the sea links between Britain and the world; had the Germans succeeded in doing so they would have won the war. They failed, however, largely because of the weaknesses in the support structure on which the naval campaign rested. While naval signals were more complex than those of the army and air force, British code breakers at Bletchly Park eventually broke into them. As a result of the Bismarck breakout, the British captured a German weather ship and a U-boat with their enigma machines and the settings for the next several months; this success gave Bletchly Park the settings of the naval enigma for the next two months. The British not only read naval signals traffic for that period, but broke into the German naval codes for the rest of the year.[105] British merchant losses, which had almost reached half a million tons per month, dropped off dramatically to only about 100,000 tons. This instance is the only case in World War II (and perhaps modern military history) where intelligence by itself proved decisive. In early 1942 the Germans added an additional enciphering wheel to their naval enigma machines; it took the British another year to break back into the naval codes. By the end of 1942 the British had again succeeded, and German failures in signals discipline contributed immeasurably to the defeat of the U-boat offensive in the climactic battles in the North Atlantic in winter and spring 1943.[106]

Conclusion

Pictures perhaps best illustrate the strengths and weaknesses of the German military in the first half of this century. There are the images of long grey lines of troops captured in France in 1940 or the Soviet Union in 1941, but these contrast with images of German cities in 1945: Cologne with its cathedral standing alone in a sea of wreckage, Berlin with its burnt out shells of buildings, or Hamburg looking like the surface of the moon.

Despite their defeats the German military performed at extraordinary levels on the battlefields in two world wars. To the end they displayed fighting power and battlefield effectiveness that their opponents found hard to match even given enormous superiority in numerical and economic strength. But down to defeat they did go; when it was all over in May 1945 they had come close to bringing the entire structure of European civilization down along with them.

It would be unfair to blame the German catastrophe entirely on the German military. The political leadership bears much of the responsibility for the disasters, as do the German people themselves for providing enthusiastic support for the regimes that dragged them over the edge. But the military is accountable for failing to analyze the strategic and political equation with honesty. As a recent article on military effectiveness has noted,

> strategic wisdom, however deserved, is more important than operational and tactical effectiveness. The best outcome, in which prewar strategic analysis helps to make force structures and operational concepts effective in wartime, is as rare in history as wise political leadership. Few "got it right" in World War I; in uniform or in mufti. . . . Military success in the earliest stages of modern war does not necessarily testify to strategic judgement. Such an assessment can only be made when the war ends.[107]

When the dust settled in both 1918 and 1945 the German military had proven themselves inept politically and incompetent strategically. All of their operational and tactical competence could not redress the strategic and political mistakes; they could only make the consequences of those mistakes more terrible for all concerned.

NOTES

1. For the clearest description of the Jena-Auerstadt disaster see David Chandler, *The Campaigns of Napoleon* (New York, 1966). For the recovery of the Prussian army after the defeat see in particular Peter Paret's two brilliant studies, *Clausewitz and the State* (Princeton, N.J., 1976) and *Yorck and the Era of Prussian Reform* (Princeton, N.J., 1966).

2. See in particular the letter from the panzer general Leo Geyer von Schweppenburg to B. H. Liddell Hart in which he notes three great intellectual influences in the curriculum at the Kriegsakademie: Frederick the Great, Napoleon, and Schlieffen. Letter from Leo Geyer von Schweppenburg to B. H. Liddell Hart, 3.8.48, King's College Archives, London, 9/24/61.

3. Holger Herwig, "The Dynamics of Necessity: German Military Policy during the First World War," in *Military Effectiveness*, vol. I, *World War I*, edited by Allan R. Millett and Williamson Murray (London, 1988), p. 81.

4. Holger Herwig, "Strategic Uncertainties of a Nation State: Prussia-Germany, 1871-1918," in the Mershon Center's "Making of Strategy" project, publication scheduled in 1993.

5. Holger Herwig, *The German Naval Officer Corps: A Social and Political History, 1890-1918* (Oxford, 1973), p. 26.

6. Quoted by Herwig in "Dynamics of Necessity," p. 79.

7. See in particular Wilhelm Deist, *The Wehrmacht and German Rearmament* (London, 1981), chapter 1.

8. Gerhard Ritter, *Staatskunst und Kriegshandwerk: Das Problem des 'Militärismus' in Deutschland*, vol. II (Munich, 1965), p. 262.

9. See Deist, *The Wehrmacht and German Rearmament*.

10. For a more detailed look at the relationship between the German rearmament effort under Hitler and the growing economic crisis, see Williamson Murray, *The Change in the European Balance of Power, The Path to Ruin, 1938-1939* (Princeton, N.J., 1985), chapter 1.

11.   On Tirpitz's designs and political effectiveness, see Holger Herwig, *'Luxury' Fleet, The German Imperial Navy, 1888-1918* (London, 1980), and Volker Berghan, *Der Tirpitz Plan: Genesis und Verfall einer innenpolitischen Krisenstrategie unter Wilhelm II* (Düsseldorf, 1971); and Paul Kennedy, "Tirpitz, England and the Second Naval Law of 1900: A Strategical Critique," *Militärwissenschaftliche Mitteilungen*, 1970.

12.   Herwig, "The Dynamics of Necessity," p. 84.

13.   For an excellent discussion of the technological advances involved in German weapons systems development see Brian Jones, *The Secret War* (London, 1978).

14.   V. R. Berghan, *Germany and the Approach of War in 1914* (London, 1973), p. 22.

15.   David Schoenbaum, *Zabern 1913, Consensus Politics in Imperial Germany* (London, 1982).

16.   See particularly Holger Herwig, "Clio Deceived, Patriotic Self-Censorship in Germany after the Great War," *International Security*, (Fall 1987).

17.   For the most thorough critique of this triumph of military influence in Germany during the war see Gerhard Ritter, *The Sword and the Sceptor*, vol. III, *The Tragedy of Statesmanship—Bethman Hollweg as War Chancellor (1914-1917)* and vol. IV, *The Reign of German Militarism and the Disaster of 1918* (Coral Gables, Fla., 1972-73).

18.   For Ludendorff's capacity in the area of tactics and military doctrine see Timothy Lupfer, *The Dynamics of Doctrine: The Changes in German Tactical Doctrine during the First World War* (Leavenworth, Kansas, 1981).

19.   There was some justification for Seeckt's reluctance because many of the troops in Bavaria showed signs of unreliability. For the army and the putsch see Harold J. Gordon, Jr., *Hitler and the Beer Hall Putsch* (Princeton, 1972).

20.   Peter Hayes, "'A Question Mark with Epaulettes'? Kurt von Schleicher and Weimar Politics," *Journal of Modern History*, 52 (March 1980).

21.   Even some of the most morally upstanding members of the eventual military opposition to Hitler, including Klaus von Stauffenberg and Ludwig Beck, were enthusiastic supporters of the Nazi movement in its early drive to power.

22.   For a discussion of the Munich crisis see Telford Taylor, *Munich, The Price of Peace* (New York, 1979). For the strategic and military aspects of the crisis see Murray, *The Change in the European Balance of Power*, chapters 5, 6, 7. For the conflict within the German leadership over the political and strategic balance throughout the crisis see Williamson Murray, "German Net Assessment in the 1930s," in Williamson Murray and Allan R. Millett, *Calculations, Net Assessment*

in the 1930s (New York, 1991).

23.   Letter from Manstein to Beck, July 1938, Beck Nachlass, Bundesarchiv/ Militärarchiv (BA/MA), N 28/3.

24.   BA/MA, n 28/3, Beck Nachlass, "Nachtrag am 19.7.38."

25.   In English see the excellent article by Jürgen Förster, "Hitler's War Aims Against the Soviet Union and the German Military Leaders," *Militärhistorik Tidschrift*, 1979; for a more complete discussion see Horst Boog, *et al.*, *Das Deutsche Reich und der Zweite Weltkrieg*, vol. IV, *Der Angriff auf die Sowjetunion* (Stuttgart, 1983), chapter 7.

26.   Ibid., pp. 1,052-53.

27.   Carl von Clausewitz, *On War*, translated and edited by Peter Paret and Michael Howard (Princeton, 1975), pp. 86-87.

28.   For the best discussion of the weaknesses involved in the Schlieffen Plan see Gerhard Ritter, *The Schlieffen Plan, Critique of a Myth* (New York, 1958).

29.   For the logistic weaknesses involved in the Schlieffen Plan see Martin van Creveld, *Supplying War, Logistics from Wallenstein to Patton* (Cambridge, 1977), chapter 4.

30.   Letter from Leo Geyer von Schweppenburg to B. H. Liddell Hart, King's College Archives, London, 9/24/61. , p. 32.

31.   The end of tour report by the future general Albert C. Wedemeyer upon his return from a two-year tour in Berlin as a student at the Kriegsakademie makes it clear that the German military education system did not expose its students to Clausewitz. See "German General Staff School," Report 15,999 dated 17/11/38 from the Military Attache Berlin, 1kb, 6/23/39, National Archives.

32.   Quoted in Herwig, "The Dynamics of Necessity," p. 40.

33.   For the best discussion of German naval attitudes see Holger Herwig, *The Politics of Frustration, The United States in German Naval Planning, 1889-1941* (Boston, 1976), pp. 225-34.

34.   For Beck's arguments see "Betrachtungen zur gegenwärtigen mil. politischen Lage," 5.5.38; "Bemerkungen zu den Ausführung des Führers am 28.5.38," 16.7.38; and Vortragsnotiz vom 29.7.38, all in BA/MA N 28/3, Nachlass Generaloberest Beck.

35.   Herwig, *The Politics of Frustration*, p. 151.

36.   Michael Geyer, *Aufrüstung oder Sicherheit, Die Reichswehr in der Krise der Machtpolitik, 1924-1936* (Wiesbaden, 1980), p. 80.

37.   Herwig, "The Dynamics of Necessity," p. 90.

38.   For the best summary of these issues in English see Herwig, *'Luxury' Fleet*.

39.   Quoted in Herwig, "The Dynamics of Necessity," p. 91.

40.   The entry of 11 August 1941 in the Halder Diary captures this miscalculation most graphically: "The whole situation makes it increasingly plain that we have underestimated the Russian colossus, who consistently prepared for war with that utterly ruthless determination so characteristic of totalitarian states. . . . At the outset of the war, we reckoned with about 200 enemy divisions. Now we have counted 360. These divisions indeed are not armed and equipped to our standards, and their tactical leadership is often poor. But there they are, and if we smash a dozen of them, the Russians simply put up another dozen." Franz Halder, *The Halder War Diary, 1939-1942* (Novato, California, 1988), p. 506.

41.   See Klaus Reinhardt, *Die Wende vor Moskau* (Stuttgart, 1972), pp. 123-71.

42.   Van Creveld, *Supplying War*, pp. 128-29.

43.   Boog, et al., *Das Deutsche Reich und der Zweite Weltkrieg*, vol. IV, p. 117.

44.   Van Creveld, *Supplying War*, pp. 153-57.

45.   See in particular Gerald Feldman, *Army, Industry, and Labor in Germany* (Princeton, 1966).

46.   See in particular Murray, *The Change in the European Balance of Power*, chapter 1.

47.   See Williamson Murray, "Force Structure, Blitzkrieg Strategy and Economic Difficulties: Nazi Grand Strategy in the 1930s," chapter 9 in this volume.

48.   For a discussion of the course and impact of German overconfidence on the mobilization of the aircraft industry see Williamson Murray, *Luftwaffe* (Baltimore, 1985), pp. 92- 104.

49.   Herwig, "The Dynamics of Necessity," p. 89.

50.   See Macgregor Knox's excellent study of this relationship in *Mussolini Unleashed* (Cambridge, 1982).

51.   Herwig, "The Dynamics of Necessity," p. 87.

52.   Walter Warlimont, *Inside Hitler's Headquarters* (New York, 1964), p. 54.

53.   British Air Ministry, *The Rise and Fall of the German Air Force, 1933-1945* (New York, 1983), pp. 101-07.

54.   Klaus Maier, et al., *Das Deutsche Reich und der Zweite Weltkrieg*, vol. II, *Die Errichtung der Hegemonie auf dem europäischen Kontinent* (Stuttgart, 1979), pp. 221-24.

55.   For the best description of army plans for "Sea Lion" see Telford Taylor, *The Breaking Wave* (New York, 1967).

56.   Meyer, et al., *Das Deutsche Reich und der Zweite Weltkrieg*, vol. II, pp. 378-79.

57.   Chief WFA, 30.6.40, "Die Weiterführung des Krieges gegen England,"

International Military Tribunal, *Trial of Major War Criminals*, vol. XXVIII, 301-03.

58.   Taylor, *Breaking Wave*, p. 216.

59.   For the best discussion of Weserübung in English see Telford Taylor, *The March of Conquest* (New York, 1958).

60.   Bradley J. Meyer, "Operational Art and the German Command System in World War I," Ph.D. Dissertation, The Ohio State University, 1988, p. 296.

61.   Robert Debs Heinl, *Dictionary of Military and Naval Quotations* (Annapolis, Maryland, 1966), p. 239.

62.   For the "Battle Of the Frontiers" see Barbara Tuchman, *The Guns of August* (New York, 1962), pp. 163-193. For the Somme see Martin Middlebrook, *The First Day on the Somme* (New York, 1972).

63.   For the best examination of these problems see G. C. Wynne, *If Germany Attacks, The Battle in Depth in the West* (London, 1940), and Timothy T. Lupfer, *The Dynamics of Doctrine: The Changes in German Tactical Doctrine During the First World War* (Leavenworth, Kansas, 1981).

64.   Erich Ludendorff, *Ludendorff's Own Story*, vol. I (New York 1919), p. 324.

65.   Ibid., p. 24.

66.   This was definitely the case with the French army even in the late periods of World War I. By early 1918 Marshal Pétain, commander of the French army, had promulgated defense in depth as the official doctrine of the army; the disastrous pummeling that the British took in March and April 1918 should only have served to reinforce the lesson that a defense in depth represented the only possible way to stop an enemy attack based on the new offensive tactics. Nevertheless, General Duchesne, commander of the Sixth Army, entirely disregarded his instructions and ordered his subordinate commanders to pack their troops into the front line, where the German offensive with its heavy artillery bombardment slaughtered them in the great offensive against Chemin des Dames. Barry Pitt, 1918, *The Last Act* (New York, 1962), pp. 143-48.

67.   See in particular the outstanding monograph by Robert Doughty, *Seeds of Disaster, The Development of French Army Doctrine 1919-1939* (Hamden, Conn., 1985).

68.   Michael Howard, *The Continental Commitment* (London, 1972), p. 32. For the best general study of the British army in the interwar period see Brian Bond, *British Military Policy between the Two World Wars* (Oxford, 1980).

69.   This, of course, resulted in much bitterness. The Nazi movement attracted much of this disaffected segment of the officer corps; Ernst Röhm and his cohorts opposed the army not just because it was a conservative organization, but because it had preferred the general staff officers to war heroes like themselves.

70.   Seeckt succeeded to the point that virtually all of the senior army commanders in World War II had at one point or another been general staff officers during their careers. Rommel was very much an exception, as most of the disparaging remarks by his colleagues in the postwar literature make clear.

71.   See chapter 10 of this collection.

72.   That this was not always the case is underlined by the performance of much of the British army in World War II. See Williamson Murray, "British Military Effectiveness in the Second World War," Millett and Murray (eds.) *Military Effectiveness*, vol. III.

73.   The exact opposite was the case in the experience of this author in the USAF during the 1960s.

74.   Barry Posen in his book *The Sources of Military Doctrine, France, Britain, and Germany Between the Wars* (Ithaca, New York, 1984) argues the opposite. Considering that Posen does not refer to a single original document, does not use any secondary sources that are in a foreign language, and has obviously not even consulted "Dissertations in Print," this represents a fundamentally flawed work that is dangerously superficial.

75.   For a more thorough exposition of these factors see chapter 10 of this work.

76.   Quoted in Jürgen Förster, "New Wine in Old Skins? The Wehrmacht and the War of 'Weltanschauung,'" in Wilhelm Deist, ed., *The German Military in the Age of Total War*, (Leamington Spa, 1985), p. 305.

77.   Quoted in Jürgen Förster, "The Dynamics of Volksgemeinschaft: The Effectiveness of German Military Establishment in the Second World War," in Millett and Murray, *Military Effectiveness*, vol. III, p. 205.

78.   See Omar Bartov, *The Eastern Front 1941-45, German Troops and the Barbarization of Warfare* (New York, 1986).

79.   For a particularly good discussion of the operational consequences of logistic failure see van Creveld, *Supplying War*, chapter IV.

80.   See the table "Die materiele Ausstattung des deutschen Ostheeres am 22 Juni 1941," in Horst Boog, *et al.*, *Das Deutsche Reich und der Zweite Weltkrieg*, vol. IV, *Der Angriff auf die Sowjetunion* (Stuttgart, 1983), pp. 186-87.

81.   Van Creveld, *Supplying War*, p. 151.

82.   Ibid., p. 153.

83.   See Reinhardt, *Die Wende vor Moskau*, pp. 123-71.

84.   For the clearest technical discussions of why and how the message traffic proved vulnerable see Gordon Welchman, *The Hut Six Story* (New York, 1982).

85.   For the message on Panzer Group West's location see PRO DEFE 3/168. KV 71781, 9.6.44, 2044Z; and KV 7225, 10.6.44, 0439Z. For the messages on the

Mortain Counterattack in August see Ralph Bennett, *Ultra in the West, The Normandy Campaign 1944-1945* (New York, 1979), pp. 114-16.

86.   For an outstanding work that unravels this enormously important contribution to the winning of World War II see David Glantz, *Soviet Military Deception in the Second World War* (London, 1982).

87.   Given the capabilities of air power in World War I, it is hardly necessary to discuss the German air force as an independent service in that conflict. The airplane represented a weapon of great promise but little more. Nevertheless, the Germans did introduce "strategic" bombing to the world in 1916.

88.   For a discussion of the issues involved in the Luftwaffe's preparation for the "strategic" bombing mission see chapter 2 For other analyses of the issues involved see Murray, *Luftwaffe*, 11-13. See also the excellent discussion in Klaus Maier, *et al.*, *Das Deutsche Reich und der Zweite Weltkrieg*, vol. II, pp. 43-49.

89.   I am indebted to Oberst Dr. Klaus Maier of the Militärgeschichtliches Forschungsamt in Freiburg for providing me with a copy of the Luftwaffe's basic doctrinal statement, "Die Luftkriegführung (Conduct of the Air War)."

90.   Edward Homze, *Arming the Luftwaffe, The Reich's Air Ministry and the German Aircraft Industry, 1919-1939* (Lincoln, 1976), pp. 223-24.

91.   Johannes Steinhoff, *Messerschmitts over Sicily* (Baltimore, 1987), pp. 59-61.

92.   U.S. Army Air Force, *Ultra and the History of the United States Strategic Air Force versus the German Air Force* (Frederick, Maryland, 1980), p. 13.

93.   David Irving, *The Rise and Fall of the Luftwaffe, The Life of Field Marshal Erhard Milch* (Boston, 1973), p. 148.

94.   Horst Boog, *Die Deutsche Luftwaffenführung, 1933-1945* (Stuttgart, 1982).

95.   Air Historical Branch (Great Britain), "Luftwaffe Strength and Serviceability Tables, August 1938-April 1945," Translation No. VII/107.

96.   Irving, *The Rise and Fall of the Luftwaffe*, p. 134.

97.   Horst Boog, "Higher Command and Leadership in the German Luftwaffe, 1933-1945," in *Airpower and Warfare*, ed. by Alfred F. Hurley and Robert C. Ehrhart (Washington, 1979), p. 145.

98.   For a translation of this intelligence study see Francis K. Mason, *Battle over Britain, A History of the German Air Assaults on Great Britain, 1917-1918 and July-December 1940 and of the Development of Britain's Air Defenses Between the World Wars* (Garden City, New York, 1969), appendix K.

99.   Basil Collier, *The Defense of the United Kingdom* (London, 1957), p. 160.

100.   Irving, *The Rise and Fall of the Luftwaffe*, p. 123.

101.   Herwig, "The Dynamics of Necessity," p. 99. 102.

Ibid., p. 99.

103.   Michael Salewski, *Die Deutsche Seekriegsleitung*, vol. I (Frankfurt am Main,

1970), p. 29.

104.   BA/MA, M/31/PG3458, Marine Kriegsadademie, Winterarbeit Kptlt Haack, 1938/1939, "Welche Wege kann die Seestrategie Englands in einem Krieg gegen Deutschland einschlagen und welche strategischen und operativen Möglichkeiten ergeben sich daraus für die deutsche Seekriegführung?"

105.   See Patrick Beesley, *Very Special Intelligence* (Garden City, N.Y., 1978).

106.   Here, however, there were other contributing factors: the addition of long-range aircraft, more anti-submarine vessels, better tactics, and new technology all contributed along with intelligence to the winning of the Battle of the Atlantic.

107.   Allan R. Millett and Williamson Murray, "Lessons of War," *The National Interest*, Winter 1988/89.

CHAPTER
TWO

THE LUFTWAFFE BEFORE
THE SECOND WORLD WAR:
A MISSION, A STRATEGY?

The title of this chapter is meant to suggest that the Luftwaffe's conception of both itself and its mission before the Second World War was quite different from the generally held historical view. The most common assumption has been that before the Second World War the Luftwaffe, refusing to heed the prophecies of Douhet, Trenchard, and Mitchell, prepared itself to support the German army as a part of a well-orchestrated strategy, known generally as "Blitzkrieg strategy."[1] As a result, the Luftwaffe supposedly never considered "strategic" bombing[2] and entered the war as a tactical air force incapable of affecting the course of the war beyond supporting the army.

Such an interpretation does not hold up, however, when one examines the documents. What has happened is that some Anglo-American historians have imposed a preconceived framework that perceives a relationship between events, tactical developments, and strategy that did not exist at the time. This process undoubtedly reflects the natural human tendency to look for order and clarity in history. Unfortunately, human affairs are inherently disorderly and irrational; the search for order may only serve to obscure and distort rather than illuminate.

Before turning to the Luftwaffe and its mission conceptualization, an examination of certain aspects of the Blitzkrieg strategy theory to place the prewar development of air power in Nazi Germany is helpful. First to note is that the criteria that determined German strategy and force structure were very different from those of the United States and Britain. Economic historians have emphasized the low level of German armaments production in the late 1930s (in relation to what Germany would produce in 1943).[3] For example, Britain had surpassed German aircraft production by 1940.[4] Supposedly this state of affairs resulted from Hitler's desire to fight short, quick wars and his refusal to impose a lower standard of living on the German working class. In other words, faced with the choice of guns or butter, he chose butter and minimized German rearmament, while concentrating German strategy and production on a Blitzkrieg strategy: the tank/aircraft team working in close cooperation to overthrow Germany's opponents one at a time.

There are several major flaws in such a thesis, of which the most important is that an entirely different economic situation determined German rearmament during the interwar period than was the case in Britain and particularly the United States. The Germans possessed few natural resources and had to import nearly every raw material that a modern war economy requires. Concurrently, they had to export finished industrial goods to cover the costs of raw materials needed by rearmament. Thus, the German economy was most vulnerable to a blockade. All of this put the Germans in a considerable bind in the late 1930s.[5]

In terms of air rearmament, it meant that in the prewar period Germany was never in a position to build a "strategic" bombing force on the scale of the British and American bomber fleets that battered Germany in 1943 and 1944. Hitler's megalomaniacal demand after Munich for a fivefold increase of the Luftwaffe was never in the cards: such a force would have required 85 percent of the world's supply of aviation gas and would have cost 60 million Reich marks, equivalent to the entire defense spending of the 1933-39 period. Most senior Luftwaffe officers recognized the demand as patently impossible.[6]

Did the Luftwaffe then turn to a tactical conception of its role and subordinate itself to a strategy of close cooperation with the army? Here the historical theory of a "Blitzkrieg strategy" supported by "Blitzkrieg tactics" breaks down. To begin with, from what we know of Hermann Göring, a man who jealously guarded his prerogatives and position within the Nazi system, is it likely that he would have deliberately subordinated the Luftwaffe, his power base, to an army that he regarded as conservative, hidebound, and politically reactionary? Admittedly, Göring's control of the Luftwaffe was at times rather loose (in 1938 he admitted to German industrialists that he saw his technical experts approximately once a week).[7] Were there then other senior Luftwaffe officers in a position to chart an independent course? Here again the answer is no. By the gradual isolation of his strong state secretary, Erhard Milch, and by the support he tendered to individuals like Ernst Udet and Hans Jeschonnek, Göring insured in the late 1930s that no one else possessed the authority or position to act independently within the air force command system.

How close air support doctrine developed within the Luftwaffe also undermines the Blitzkrieg theory. The technique and tactical side of close air support emerged in an *ad hoc* manner and with even less support from the air force high command than the armored forces received from the army.[8] In other words, the doctrine and tactics of the close support mission evolved without a push from above,[9] but rather from German experience in Spain and in particular from Wolfram von Richtofen's drive and imagination. Richtofen himself arrived in Spain out of favor with the air ministry in Berlin. His conception of air war upon arrival was not substantially different from that of most other Luftwaffe officers at that time; close air support for the army ranked at the bottom of his priorities. However, once in his position as chief of staff of the German air units in Spain, Richtofen recognized that little common ground stood between theories of air power and Spanish political realities. The stalemate on the ground, the lack of suitable strategic targets, and the great Nationalist weakness in artillery led Richtofen to consider using the Condor Legion to support Franco's offensive against Bilbao.[10]

Against considerable opposition and without official sanction, Richtofen developed the technique and tactics of close air support for ground forces in offensive operations.[11] None of the elements required for such operations existed before the offensive against the Basque Republic. Air to ground communications were crude, and the technical means for cooperation between ground and air units (even on the simplest level, not to mention radio) either did not exist or had not yet been employed. By the time Richtofen was through, the Germans had recognized the need for cooperative planning between ground and air, had established communication links and recognition devices, and had detailed Luftwaffe liaison officers to serve directly with frontline units.[12]

Nevertheless, even upon the return of the Condor Legion to Germany, close air support remained a small percentage of the Luftwaffe's force structure. In August 1939, aircraft assigned directly to this mission made up less than 15 percent of German combat aircraft.[13] This was, of course, a significant number, especially when compared to the air forces of other nations, yet it suggests that close air support was not *the* chief mission of the Luftwaffe.

One aspect of German aircraft development must be addressed before turning to the Luftwaffe's own conception of its strategy and mission in a future war. Beginning in 1937, Ernst Udet, head of research and development as well as production, demanded that German designers incorporate into future bomber aircraft the capability of dive bombing. From an engineering standpoint, such a requirement was unrealistic and the failures of the Ju 88 and He 177 programs were in part due to this requirement. That Udet's demand reflected a desire that the bombers serve as close air support for the army is not supported by evidence. Experience in high altitude bombing in Spain showed the problems involved in hitting targets accurately, and Udet seems to have believed that a dive bomber capability would solve these problems.[14] The low capacity of Germany's munitions industry (1938 maximum capacity was less than 30 percent of World War I production)[15] suggested the possibility that such a capability would cut down munitions wastage and thus mitigate possible ammunition shortages.

The development of Luftwaffe doctrine before the war is another consideration in understanding its preparation for war. Two dominant themes in Luftwaffe thinking in the prewar period still prevailed at the outbreak of the Second World War. The more sophisticated theme came from the statements and work of General Walther Wever, the Luftwaffe's first chief of staff. Wever was not an unabashed advocate of "strategic" bombing as some historians have claimed.[16] Rather he articulated a broadly based view of air power in which the Luftwaffe would cooperate with the other services to achieve an overall strategic result. The conditions of the general situation would guide the Luftwaffe's employment. Its contribution to victory could involve direct support of the army or navy, or in other circumstances the destruction of an enemy's resources and armament industries through "strategic" bombing.[17]

The formulation of German air doctrine appearing in 1935, "The Conduct of the Air War" (*Die Luftkriegführung*), best sums up Wever's approach.[18] As with most German doctrine, this manual was a clear, concise statement of principles. Its aim was not to restrict or to dogmatize but rather to give commanders the widest latitude and maximum flexibility. It argued that the Luftwaffe must exist within the larger framework of national strategy: "The nature of the enemy, the time of year, the structure of his land, the character of his people as well as one's own military capabilities" would determine the employment of air power.

"The Conduct of the Air War" emphasized that air power should not be employed in a piecemeal fashion. In a future war, it suggested, the Luftwaffe would probably not be able to make a clear separation between combating the enemy's air force and the necessity to support the army and navy. Unlike most contemporary air power theories, this statement grasped that air superiority would be a most elusive goal. It recognized that "strategic" bombing of the enemy's industrial and economic resources might take too long to achieve a decision. Wever and "The Conduct of the Air War" clearly did not eliminate "strategic" bombing as a proper employment for aircraft, but rather suggested that it represented only one of several possible missions.

Wever's death in 1936 had a shattering impact on the Luftwaffe's command structure, and none of his successors as chief of staff before (or during) the war possessed his strength of personality or his strategic breadth and clarity.

Wever's approach to air power was not universal within the Luftwaffe. Like many of their Allied peers, much of the Luftwaffe's officer corps saw "strategic" bombing as the only proper role for air power. As early as May 1933 when the Germans were first formulating plans for their new air force, Dr. Robert Knauss of the air ministry urged that the decisive element in the new Luftwaffe be a fleet of 400 four-engine bombers. Knauss argued that such a fleet offered the possibility to attack not only the enemy's industrial production, but population centers as well, and went so far as to suggest that "the terrorization of the main cities and industrial regions of the enemy by bombing would lead that much more quickly to a collapse of morale, the weaker the national character of his people is and the more that social and political rifts cleave his society." Naturally, Knauss assumed that the totalitarian society of the new Germany would prove better able to endure bombing attacks than would the fractured, democratic societies of Britain and France.[19]

Knauss' argument reflected current thinking in those military circles, influenced by Ludendorff, who believed that the next war would be total, involving civilian society as well as the military.[20] Knauss also reflected Nazi ideology and a belief that the Nazi *Volksgemeinschaft* would strengthen and unite the German nation.

Knauss himself left the air ministry to become the commandant of the air war college at Gatow. At the college, the emphasis reflected Knauss' beliefs and rested solidly on "strategic" air warfare through to the outbreak of war. Few lectures at the college discussed tactical cooperation with the army or the broader application of air power.[21]

Similarly, German military journals emphasized "strategic" air attacks. The prestigious *Militärwissenschaftliche Rundschau*, the war ministry's new journal, published several discussions of the future employment of air power that also emphasized the role of "strategic" bombing to the exclusion of all else. One author noted that European countries were increasingly leaning toward bomber forces as the heart of their air forces. The maneuverability and technical capabilities of the new generation of bombers were such that "already in today's circumstances the bomber offensive would be as unstoppable as the flight of a shell."[22]

Major Herhudt von Rohden, eventually head of the general staff's historical section, argued that unlike the army and navy the air force was in a position to attack the enemy in depth and to launch immediately "destructive attacks against the economic resources of an enemy from all directions." Von Rohden stressed that the Luftwaffe should not serve as an auxiliary to the other services. Interservice cooperation did not mean dividing up the Luftwaffe and parceling its personnel and material out to support ground or naval tactical problems. Rather, interservice cooperation meant using one's air force "in a unified and massed 'strategic' air war." On that fundamental task alone could the air force provide the best long-range support for the other services.[23]

Such attitudes found an echo in the operational side of the Luftwaffe. General Hellmuth Felmy, commander of *Luftflotte* 2 (Air Fleet 2) and charged in the late 1930s with the conduct of a future air war against Britain, speculated in May 1939 on the possibilities offered by terror bombing attacks against London. Among other things, Felmy emphasized the psychological impact of such an offensive. For support he pointed to the fact that the mere threat of air attacks had led to a high degree of war hysteria in Britain in fall 1938. Should war occur, he suggested that the Third Reich take full advantage of such fear rather than hesitate, as the Kaiser's government had done during World War I.[24] The Fifth Section (intelligence) of the Luftwaffe's general staff articulated similar views shortly before the war. It noted that, compared with other European air forces, the Luftwaffe was the best prepared for general air warfare:

> Germany is on the basis of all reports the only state that in respect to equipment, organization, tactics and leadership has advanced to a total conception of the preparation and leadership of an offensive as well as defensive air war. The fact indicates a general advance in military preparedness and with it a strengthening of the whole military situation.[25]

Not unnaturally, given the atmosphere, the higher leadership of the Nazi state found the air weapon most attractive. While Göring never found serious thinking on strategic matters much to his taste, he certainly liked the emphasis on his Luftwaffe as a decisive, war-winning force. Hitler, who interfered little in the formulation of tactics and doctrine before the war,[26] was most impressed with the air weapon. Indeed, his sensibilities were much attracted by the *Schrechlichkeit* (frightfulness) that the aircraft had seemingly introduced into warfare.[27] In August 1939 he cited Germany's air superiority as one of the critical factors supporting his willingness to risk war in that year: ". . . as air superiority is undoubtedly on our side, I do not shrink from solving the eastern question even at the risk of complications in the West."[28]

On 1 September 1939, Germany's attack on Poland precipitated the Second World War. In that campaign, German armored doctrine, particularly the deep penetration drive of the Tenth Army, came into its own. Contributing to that success was the Luftwaffe's close air support, mostly consisting of "Stukas." The bulk of the Luftwaffe however, was not directly involved in supporting the army; rather it carried out wide-ranging attacks aimed first at eliminating the Polish air force as a military factor and then at paralyzing Poland's transportation system. Interestingly, the Germans had initially considered launching a massive, all-out attack on the military installations and armament factories of Warsaw to break Polish resistance at the start of the campaign. Bad weather, however, prevented the launching of such attacks.[29] One must also note that at the campaign's conclusion the Luftwaffe launched massive air assaults against Warsaw. While these raids aimed to destroy military targets, the Germans were not averse to the collateral damage of such attacks.

Some historians have tended to see the aftermath of Poland and the build up to the 1940 campaign in the west as a continuum in which the German military and Hitler planned for exactly the same type of offensive that they had carried out so successfully in Poland.[30] It was not. In fact, there was an inordinate amount of squabbling between Hitler and the generals about what strategy Germany should follow. Hitler demanded an immediate offensive in the west, but the generals simply had no idea of how they could possibly beat the French.[31] As late as March 1940 the army high command was still opposing a major offensive through the Ardennes.[32] The reasons for Hitler's initial strategy of launching the Wehrmacht west in fall 1939 have also caused considerable debate. Disregarding the economic factors pushing in that direction,[33] Hitler seems to have believed that a limited campaign in which his forces seized Belgium and Holland would give a twofold advantage: the first "as a base for the successful prosecution of the air and sea war against England," and the second as a buffer to protect "the economically vital Ruhr."[34] Thus, air strategy seems to have played the major role in his calculations.

Echoing the Führer's sentiments, the Luftwaffe's chief of intelligence, "Beppo" Schmid, argued in late November 1939 for an exclusive air strategy in which the Wehrmacht would launch no operations against the French, but rather the entire Luftwaffe with whatever help the navy could provide, would concentrate against English imports. The major effect would emphasize English ports and docks, and Schmid noted that "should the enemy resort to 'terror' measures, for example to attack our towns in western Germany, here again similar operations could be carried out with even greater effect, due to the greater density of population of London and the big industrial centers."[35]

Hitler eventually settled on the great armored drive through the Ardennes. In that campaign the Luftwaffe's strategy remained closely tied to the overall demands of the effort on the ground. German close support aircraft (a small percentage of the overall bomber force) pro-

vided invaluable support for the armored force, while the bulk of the
Luftwaffe attacked Allied air forces and the transportation network. The
former prevented enemy air forces from intervening effectively against
the armored drive, while the latter contributed to the paralysis of the
Allied command system.[36] The effectiveness of the air campaign as a
whole raises an interesting question as to what exactly the strategic use
of air power is. It has often been exclusively identified with "strategic"
bombing. But in this case, the great German victory over France—a
victory that destroyed the European balance of power—did not the
Luftwaffe's mission and strategy, closely integrated into overall German
strategy, reflect a more strategic employment of air power than if the
Germans had launched their air resources against industrial targets or
population centers in the depth of France?

Several features of prevailing opinion on Luftwaffe doctrine before
the war deserve reconsideration. The first and most obvious is that the
picture of a Luftwaffe closely tied to the army's coattails is no longer
tenable. Many—if not a majority of—officers within the Luftwaffe believed that "strategic" bombing was their chief mission, and that in such
a role they would win the next war. They probably did not consider the
twin-engine aircraft at their disposal in 1937 and 1938 sufficient for a
campaign against Britain, Russia, or the United States. But within the
context of Central Europe were not such aircraft adequate for attacking
Warsaw, Prague, and Paris? Most Germans thought so, and certainly
the leaders of the French and British air forces would have agreed.

For the long run the Luftwaffe had begun work on a four-engine
bomber for more distant targets. The failure to complete that program
reflected more the failures of German engine development and the inadequacies of Luftwaffe leadership than a renunciation of "strategic"
bombing.[37] Like most of their contemporaries in other air forces,
Luftwaffe officers considerably overestimated the possibilities and potential of "strategic" air war, both in terms of industrial damage and its
impact on morale. This was neither surprising nor unique since there
was so little empirical evidence on which to base predictions. The preva-
lence of such attitudes within the Luftwaffe's officer corps does help to
explain the terror bombing of Rotterdam as well as the seemingly casual
shift from an air superiority strategy to a direct attack on London during
the Battle of Britain. Moreover, in their approach to "strategic" bombing,
the Germans showed greater awareness of the difficulties in finding and
hitting targets at night or in bad weather than did other air forces, and

their preparations in developing blind bombing devices like *Knickebein* were considerably in advance of the RAF.

On the other hand, there were factors pushing the Germans towards a broader conception of air power. As discussed earlier, economic factors placed severe limits on the prewar force structure of the Luftwaffe. But even more important was Germany's geographic position in the heart of Europe. Unlike British and American air strategists, German air strategists faced the prospect of a land battle from the moment that a war began. Thus, in the mid 1930s it would be of little benefit to the Reich to launch "strategic" bombing attacks against Paris, Warsaw, or Prague at the same time that enemy ground forces seized the Rhineland, or Silesia. In the late 1930s, when Germany had more scope for offensive operations, Germany had to win the first land battles to gain the resources to sustain a long war. If Germany did not win those first battles, the war was irrevocably lost. The Americans and British, however, secure behind their ocean barriers, could be more philosophical about the course of the war on the Continent. To them the loss of the first confrontations on land would not be immediately decisive, and they could thus afford to think in terms of a different air war than did those on the Continent.

In reality, the Luftwaffe's doctrine and conception of its role in German strategy was a mixture of the above two tendencies. When war came, German airmen found themselves in quite different strategic circumstances than they had originally envisioned. Unfortunately for the West, Wever's more broadly based approach and the resulting greater flexibility in Luftwaffe thought and doctrine, corresponded more closely to actual aircraft capabilities in the late 1930s than did the almost exclusively "strategic bombing" dogma of the RAF or the US Army Air Corps. The real war of 1939 and 1940 was not the war for which most of the Luftwaffe had prepared, but it was a war—in its initial stages—to which the Luftwaffe could and did adapt, and to which the Luftwaffe could in the final analysis apply air power in cooperation with the army to gain a devastating strategic victory.

NOTES

1.   For the generally held assumption that the German air force was the "handmaiden" of the army see among others, Herbert M. Mason, Jr., *The Rise of the Luftwaffe* (New York, 1973), p. 215; Peter Calvocoressi and Guy Wint, *Total War* (London, 1972); Denis Richards, *The Royal Air Force, 1939-1945* (London, 1953), p. 152; Basil Collier, *The Defence of the United Kingdom* (London, 1957), p. 121; Telford Taylor, *The Breaking Wave* (New York, 1967), p. 83; and surprisingly, Charles Webster and Noble Frankland, *The Strategic Air Offensive Against Germany*, vol. I, *Preparation* (London, 1961), p. 125. For the theory of "Blitzkrieg strategy" see Larry Addington, *The Blitzkrieg Era and the German General Staff* (New Brunswick, 1971); and Alan Milward, *The German Economy at War* (1965).

2.   For the purposes of this paper the use of the word "strategic" indicates the application of air power in an independent role to strike an enemy's industry, or morale, or both.

3.   Above all see Burton Klein, *Germany's Economic Preparations for War* (Cambridge, Mass., 1959).

4.   Richard Overy, "The German Pre-war Aircraft Production Plan: November 1936-April 1939," *English Historical Review*, 1975, p. 796.

5.   For a fuller description of the German economy and the problems of rearmament see Williamson Murray, *The Change in the European Balance of Power, 1938-1939*, (Princeton, 1985) chapter 5.

6.   Edward Homze, *Arming the Luftwaffe: The Reich Air Ministry and the German Aircraft Industry, 1919-1939* (Lincoln, 1976), pp. 223-24.

7.   International Military Tribunal (IMT), *Trial of Major War Criminals* (*TWMC*), vol. XXXVIII, Doc. 140-R.

8.   For a full discussion of the development of armored doctrine see Murray, *The Change in the European Balance of Power*, chapter 1.

9.   "Lehren aus dem Feldzug in Spanien, Einsatz von Schlachtfliegern," aus einer Studie der 8. Abt. des Generalstabes aus dem Jahre 1944; Hans Hennig Freiherr von Beust, "Die deutsche Luftwaffe im spanischen Krieg," 2 Oct. 1956, p. 162, Albert Simpson Historical Research Center: Karlsruhe Collection (ASHRC: KC), K 113.302.

10.   Conversation with Generalmajor a.D. Hans W. Asmus, Baden Baden 7 and 8 Nov 1980, and letter from General Asmus, 6 Feb 1981.

11.   Air Ministry, *The Rise and Fall of the German Air Force (1933-1945)*, issued by the Air Ministry (ACAS) (London, 1948), pp. 16-17.

12.   "Lehren aus dem Feldzug in Spanien, Einsatz von Schlachtfliegern," aus einer Studie der 8. Abt. des Generalstabes aus dem Jahre 1944; Hans Hennig

Freiherr von Beust, "Die deutsche Luftwaffe im spanischen Krieg," 2 Oct. 1956, p. 162, ASHRC: KC, K 113.302.

13.   Figures are drawn from 8 Aug. 1939, and come from "Luftwaffe Strength and Serviceability Tables, Aug. 1938-April 1945," Air Historical Branch, Translation no. VII/107, 6.

14.   See particularly Edward L. Homze, "The Luftwaffe's Failure to Develop a Heavy Bomber before World War," *Aerospace Historian* (March 1977); and David Irving, *The Rise and Fall of the Luftwaffe* (Boston, 1973), p. 65.

15.   Reichsstelle für Wirtschaftsausbau, Berlin 7 Jan. 1939, "Verzögerung im Schnell plan vom 13.8.38, durch verringerte Stahlzuteilung," National Archives and Records Service (NARS) T-71/110/613255.

16.   Among others see Calvocoressi and Wint, *Total War*, pp. 133 and 492.

17.   See in particular Wever's lecture to the Air War College 1 Nov. 1935, "Vortag des Generalmajor Wever bei Eröffnung der Luftkriegsakademie und Lufttechnischen Akademie im Berlin-Gatow am 1. Nov. 1935," *Die Luftwaffe*, 1936.

18.   "Die Luftkriegführung," Berlin, 1935; copy made available to me by Oberstleutnant Dr. Klaus Maier of the Militärgeschichtliches Forschungsamt, Freiburg.

19.   Bernhard Heimann and Joachim Schunke, "Eine geheime Denkschrift zur Luftkreiskonzeption Hitler-Deutschlands vom Mai 1933," *Zeitschrift für Militärgeschichte*, vol. III (1964), pp. 72-86.

20.   For discussion of Ludendorff's conception of "total war" see Hans Speier, "Ludendorff: The German Concept of Total War," in *Makers of Modern Strategy*, ed. by Edward Mead Earle (Princeton, 1943). I would also like to thank Oberstleutnant Dr. Klaus Maier of the Militärgeschichtliches Forschungsamt, Freiburg, for making available to me his lecture given at the Air War College, Maxwell Air Force Base, in September 1980 that clarified the connection between the Ludendorff school and the development of Luftwaffe doctrine.

21.   Air Ministry, *The Rise and Fall of the German Air Force*, p. 42.

22.   Major Bartz, "Kriegsflugzeuge ihre Aufgaben und Leistung," *Militärwissenschaftliche Rundschau* (1936), p. 210.

23.   Major Herhudt von Rohden, "Betrachtungen über den Luftkrieg," Part I, *Militärwissenschaftliche Rundschau* (1937), pp. 198-200.

24.   BA/MA RL 7/42, RL 7/43, Luftflottenkommando 2., Führungsabteilung, Nr. 7093/39, 13 May 1939, "Schlussbesprechung des Planspieles 1939."

25.   Klaus A. Maier, Horst Rohde, Bernd Stegemann, and Hans Umbreit, *Das deutsche Reich und der Zweite Weltkrieg*, vol. II, *Die Errichtung der Hegemonie auf dem Europäischen Kontinent* (Stuttgart, 1979), p. 44. For a further example of

the emphasis in the Luftwaffe on waging an independent, "strategic" air war consult Chef des Organizationsstabes im Generalstab der Luftwaffe Nr 50/38 Chefsache, An den Chef des Generalstabes der Luftwaffe, "Organizationsstudie 1950," NARS T-971/36/0002 for the kind of air war favored by senior staff officers.

26.   See particularly Friedrich Hossbach, *Zwischen Wehrmacht und Hitler* (Hanover, 1949), p. 39. This of course presents a major problem for those who advocate the theory of "Blitzkrieg strategy."

27.   On the attraction that the air weapon had for Hitler as a terror weapon see Richard J. Overy, "Hitler and Air Strategy," *Journal of Contemporary History* vol. 15 (1980).

28.   *Documents on German Foreign Policy*, Series D, vol. VII, Doc. no. 192, 22 Aug. 1939.

29.   "The Luftwaffe in Poland," a study produced by the German Historical Branch (8th Abteilung) 11 July 1944. AHB Translation no. VII/33.

30.   For the most recent restatement of Hitler's support of the "Blitzkrieg strategy" see Norman Stone's *Hitler* (Boston, 1980), pp. 99-100.

31.   For a full discussion of this squabbling see Maier, *et al.*, *Das deutsche Reich und der Zweite Weltkrieg*, vol. II; and Hans-Adolf Jacobsen, *Fall Gelb: Der Kampf um den deutschen Operationsplan zur Westoffensive* (Wiesbaden, 1957).

32.   Heinz Guderian, *Panzer Leader* (New York, 1952), pp. 90-92.

33.   See the discussion in Murray, *The Change in the European Balance of Power, 1938-1939*, pp. 102-7.

34.   H. R. Trevor-Roper, ed., *Blitzkrieg to Defeat, Hitler's War Directives, 1939-1945* (New York, 1964), Directive no. 6, 9 Oct. 1939, p. 13.

35.   "Proposal for the Conduct of Air Warfare against Britain," made by General Schmid of the German Air Force Operations Staff (intelligence), 22 Nov. 1939. AHB, Translation no. VII/30.

36.   There are two areas where the Luftwaffe might be said to have stepped beyond this framework of overall German strategy: Rotterdam and Dunkirk. In the latter case might not Göring's assurance to Hitler that the Luftwaffe could destroy the British evacuation at Dunkirk reflect a desire to prevent the Luftwaffe being identified too closely with cooperation with the other services?

37.   For an outstanding discussion of why the Luftwaffe did not possess a heavy bomber see Edward L. Homze, "The Luftwaffe's Failure to Develop a Heavy Bomber before World War II," *Aerospace Historian* (March 1977).

CHAPTER THREE

GERMAN AIR POWER
AND THE MUNICH CRISIS

One of the most persistent myths in postwar historical literature has been that Neville Chamberlain saved Britain at Munich from the prospect of immediate defeat at the hands of a German air offensive and thus won the time necessary for the Royal Air Force (RAF) to win the Battle of Britain.

Supposedly the year's grace allowed the RAF to re-equip fighter squadrons with Hurricanes and Spitfires and extend the radar stations to cover the whole of the British Isles.[1] The key question which must be posed, even acknowledging the dreadful state of British air defenses, and which has not been answered previously, is whether Germany was actually in a position in 1938 to launch a "strategic" bombing offensive against Great Britain. To answer such a question it is necessary to establish the actual balance of air power in 1938 by comparing the RAF and the Luftwaffe and to discuss in detail Luftwaffe training and support services and strategic and tactical planning.

Striking similarities existed in the development of these air forces in the late 1930s. Both air staffs had to grapple with rapid expansion, a new generation of aircraft, crew training, supply and maintenance—all on a scale far exceeding anything previously experienced. If anything these problems were of such magnitude as to make the respective air forces barely fit for combat. In late 1937 the British air staff reported:

The above picture of our state of readiness for war discloses many unsatisfactory circumstances, but it must be realized that the RAF is now in the midst of a most difficult stage of transition. We are in the process of expanding from a small force which mainly consists of light single-engined aircraft to a large force equipped with aircraft which are in many cases multi-engined and all of which have a high performance and completely different characteristics from those on which the bulk of the air force has been trained. Our squadrons are manned largely by personnel who—though enthusiastic and efficient—are yet lacking in experience and higher training.[2]

A German "after action" report on the Czech crisis indicates the same situation existed in the Luftwaffe in 1938:

In the last months the following special measures had to be carried through at the same time: 1) equipment of many new units, 2) rearmament of numerous units, 3) early overhaul of about sixty percent of the front-line aircraft, 4) replacement of spare parts, 5) rebuilding of numerous aircraft in the supply depots, 6) rearmament of many aircraft, 7) accelerated introduction of overhauled motor models . . . 8) establishment of four new air groups and one new airfield . . . 10) preparation and resupply of mobilization supplies, corresponding to the newly established units, rearmed units, and transferred units. . . . The compression of these tasks into a very short time span once more and in clear fashion pointed out the known lack of readiness in maintenance of flying equipment as well as in technical personnel.[3]

Neither air arm was prepared for the war that would come. The RAF envisioned future war as a surgical operation carried out by opposing air forces on populations hundreds of miles away. For the Luftwaffe, air war was more involved with interservice cooperation, but it also thought war would be quick.

The reality of air war in World War II did not fulfill either expectation. Its strategy resembled that of the First World War, this time through attrition came in terms of aircraft, bombs, numbers of crews, training programs, fuel supplies, and munitions production. Success was measured in drops of percentage points in bomber losses rather than yards gained. One commentator has pointed out that

> despite the illusions of its protagonists of pre-war days, the air war during the Second World War . . . was attrition war. It did not supplant the operations of conventional forces; it complemented them. Victory went to the air forces with the greatest depth, the greatest balance, the greatest flexibility in employment. The result was an air strategy completely unforeseen by air commanders.[4]

Neither air force was ready to fight anything resembling such a war in the late 1930s, for they did not have the skills, tactical training, or depth required. The British official history of the strategic bombing offensive puts the basic problem of air war succinctly:

> Air superiority is not simply a question of being able to use an air force. It is a question of being able to use it effectively. From the point of view of the bombers, for example, it is not simply a question of getting through. It is a question of getting through and doing effective damage.[5]

The Royal Air Force was not ready for war in September 1938; its rearmament program had made little progress, re-equipment of fighter squadrons was only beginning, and Bomber Command had no modern air-craft in production. As a contemporary noted, "in a word, expansion caught the Air Ministry napping,and the attempt to expand, side by side with the getting up-to-date technically, resulted in confusion and muddle so that no expansion program—and they succeeded each other rapidly—was ever punctually or completely carried out."[6] When Bomber Command mobilized in September 1938, only ten out of forty-two squadrons possessed what at that time passed for heavy bombers. Reserve aircraft numbered only 10 percent of frontline aircraft, while barely 200 out of 2,500 pilots were fully "operationally ready." Many aircraft had no turrets, and spare parts were in such short supply that Bomber Command had to cannibalize some squadrons to provide them. By RAF peacetime standards less than 50 percent of the force was com-bat ready.[7]

Fighter Command was in scarcely better shape. Instead of the fifty squadrons considered the minimum for Britain's air defense, twenty-nine were mobilized, with only five of these possessing Hurricanes and none having Spitfires.[8] The Hurricanes could not operate at high al-titudes because their guns did not have the warmers required for firing above 15,000 feet. The remaining squadrons possessed obsolete Gladiators, Furys, Gauntlets, and Demons.[9] As of 1 October 1938, British first-line strength was 1,606 machines with only 412 aircraft in reserve. British estimates for France were 1,454 first-line aircraft with 730 in reserve; and Germany, 3,200 first-line with 2,400 in reserve.[10] In fact the Germans had 3,307 aircraft, including transports, but almost no aircraft in reserve.[11] The anti-aircraft situation was even less encourag-ing. The War Office provided the following table (Table 2.1) on antiair-craft guns in its post-Munich review.[12]

TABLE 2.1

Availability of Anti-aircraft Weapons

Type	Approved Program	Available	Actually Deployed
3 in.	320	298	269
3.7 in.	352	44	44
2 pdr barrels	992	50	49
Naval 3 in.	—	96	95
Search Lights	4,128	1,430	1,280

Germany was no better prepared than Britain but for different reasons. German doctrine saw "strategic" bombing operations as but one of four major air missions including air superiority, battlefield interdiction, and close air support. It stressed that the air force would be part of a team rather than a service with a wholly independent mission. Moreover, German airmen doubted whether "strategic" bombing by itself could achieve a decisive result by destroying industry or terrorizing civilians partly because of their experience in Spain.[13] Development of "strategic" bombing capability was also hindered by technical problems, rather than by conscious decisions. In the other major fields of air power, such as close air support, interdiction and reconnaissance, the Luftwaffe would be much better prepared for the coming war.

The Germans approached the question of "strategic" bombing more skeptically than the British. A partial explanation lies in the German experience in Spain. Terror bombing had, for the most part, a counterproductive effect. Captain Heye of the *Seekriegsleitung* reported the following on return from Spain in July 1938:

Disregarding the great military success accompanying use of the Luftwaffe for the immediate support of army operations, one gets the impression that our attacks on objects of little military importance, through which in most cases many women and children . . . were hit, are not a suitable means to break the resistance of the opponent. They seem far more suited to strengthening the resistance . . . Doubtless the memory of the air attack on Guernica by the [Condor] Legion still today produces an after-effect in the population of the Basques, who earlier were thoroughly friendly to Germany and in no manner communistic.[14]

The strongest element, however, in the Luftwaffe's unwillingness to place the emphasis in its armament program on "strategic" bombing lay in a realistic appraisal of Germany's geographic position on the continent. The British might have the luxury to speculate about "strategic" bombing when others were going to fight the land battles, but for Germany, whatever the circumstances, war meant land war as well as air war. Thus, "strategic" bombing had less relevance to a Germany threatened by enemy ground attacks. It would do the Germans little good to carry out extensive long-range bombing on industrial and population centers if the Rhineland, the Ruhr, and Silesia were to fall to enemy ground attacks. Even when Germany was no longer threatened by land operations, other factors constrained strategic options. A 1939 report warned that a bombing offensive against Britain would open up western Germany to unlimited air attacks and make the launching of a land offensive in the west extremely difficult. Moreover, the use of fuel and munitions for an air offensive would severely restrict supplies available for ground operations.[15]

For most of 1938 the Luftwaffe was involved in exchanging its first-generation aircraft for those with which it would fight most of the coming war. Fighter squadrons replaced their Arado Ar 68s, which were biplanes, with Bf 109s, but by the autumn of 1938 there were no more than 500 Bf 109s in the regular fighter squadrons.[16] Moreover, the transition program to Bf 109s led to a high accident rate in newly converted squadrons.[17] The two bombers in production, the Do 17 and He 111, were twin-engine aircraft, which possessed neither the speed nor bomb-carrying capacity to be "strategic" bombers. Their defensive armament was insignificant. Early models of the He 111 carried a bomb load of 500 kilograms, and while London was within their range, they could barely reach the industrial regions of the midlands from bases in western Germany. The Do 17 had an even shorter range. Its first production models could reach no further than the London metropolitan area with a 500 kilogram bomb load.[18]

Bombing attacks launched from German soil would not have had fighter escort: even when based on Pas de Calais in 1940, the Bf 109s hardly had sufficient range to stay with the bombers over London. The Ju 88, supposedly a significant advance in bomber construction, was not scheduled to begin production until April 1939 and would not reach full production until 1940.[19] In August 1938, most ground attack squadrons still had He 123s, which could carry four 50 kg bombs, and He 45s, which could carry eighteen to twenty-four 10 kg bombs.[20] In view of the scarcity of raw materials, the German aircraft industry would have been hard put to maintain its rate of production if war had broken out over Czechoslovakia.

In numbers, the Luftwaffe mustered just over 3,000 aircraft at the end of September 1938. The breakdown consisted of 1,128 bombers (none of which were Ju 88s), 773 fighters, 513 reconnaissance aircraft, 226 dive bombers, 195 ground attack aircraft, 164 naval support aircraft, and 308 transports. In May 1940, shortly before the invasion of France, Luftwaffe strength was in excess of 5,000: 666 reconnaissance aircraft, 1,736 fighters, 1,758 bombers, 417 dive bombers, 49 ground attack aircraft, 241 coastal aircraft, and 531 transport aircraft.[21] The two-year difference in strength and quality is striking. Introduction of a new generation of aircraft brought with it considerable problems in air crew training, maintenance, and supply. Throughout the summer of 1938 most of the Luftwaffe experienced a high accident rate. Particularly in Bf 109 squadrons, this resulted from troubles in aircrew training in models that were far more sophisticated than anything the pilots had previously handled. The Bf 109 with its narrow undercarriage presented fighter pilots with an especially difficult transition problem. Table 2.2 indicates how unready the Luftwaffe was to fight an air war in 1938.[22]

Table 2.2

AIR CREW STATE OF READINESS, AUGUST 1938

Type of Aircraft	Crew Training Status Authorized		
	Authorized number of crews	Fully operational	Partially operational
Strategic Reconnaissance	228	84	57
Tactical Reconnaissance	397	183	128
Fighter	938	537	364
Bomber	1,409	378	411
Dive bomber	300	80	123
Ground attack	195	89	11
Transport	117	10	17
Coastal and navy	230	71	34
Total	3,714	1,432	1,145

Third Air Force (*Luftflotte 3*) reported that two factors contributed to the relatively high number of partially operational crews (*bedingt einsatzfähig*); bomber crews who were not fully operational were not rated for instrument flying, while Stuka crews generally lacked fully trained radio operators and machine gunners.[23]

The Luftwaffe's "in commission" rate for the period 1 August 1938 to 8 December 1938 indicates the extent of supply and maintenance problems. There were significant improvements in September percentages, but this resulted from a deliberate reduction of flying and training time as the invasion of Czechoslovakia approached. Third Air Force's "after action" report on the Czech crisis admitted that the "in commission" rate had been brought to a high level by carefully planned measures, but that losses as well as the heavy demands of combat operations would have quickly lowered these rates. Moreover, its units did not possess adequate reserves of spare parts to support even normal flying.[24] By December 1938, "in commission" rates had fallen considerably.

While it is hard to estimate how much strain the Luftwaffe could have sustained before overall combat effectiveness would have suffered, August "in commission" rates, as shown in table 2.3, could not have given the Luftwaffe staff much encouragement:[25]

Table 2.3

LUFTWAFFE "IN COMMISSION" RATES IN PERCENTAGES
1 AUGUST 1938 TO 8 DECEMBER 1938

	1 Aug	15 Aug	5 Sep	12 Sep	19 Sep	26 Sep	8 Oct	8 Dec
Bombers	49	58	76	84	89	90	90	78
Fighters	70	78	89	88	93	95	90	78
Overall	57	64	79	83	90	94	92	79

As already mentioned, the high September percentages do not indicate that the Luftwaffe had solved its maintenance problems. The chief of supply services concluded that maintenance and supply during the Czech crisis had led to severe consequences:

a) a constant and, for the first line aircraft, complete lack of reserves both as accident replacements and for mobilization; b) a weakening of the aircraft inventory in the training schools in favor of regular units; c) a lack of necessary reserve engines, supplies for the timely equipment of airfields, supply services and depots both for peacetime needs as well as for mobilization.[26]

The Luftwaffe's chronic supply weaknesses in World War II contradicted its own philosophy of air power. The Luftwaffe staff believed that a flying unit was not combat ready unless it possessed modern reliable aircraft, backed up by a first-class maintenance organization and by a supply system that guaranteed adequate numbers of replacement aircraft and reserves of spare parts. The chief of the Luftwaffe's supply branch reported in an "after action" report on the Czech crisis that "these three requirements were not met."[27] There were no reserve aircraft to replace combat losses because the Luftwaffe had devoted its entire production to supplying frontline units with new aircraft or to equipping newly established squadrons.

The situation was the same in all categories of spare parts. The number of aircraft engines in maintenance and supply depots represented only 4 to 5 percent of total engines in service, an incredibly low figure. The number of built up engines was correspondingly lower. Considering the number of airframes and engines in the supply pipelines, the supply staff doubted whether the Luftwaffe could have fought for more than four weeks.[28] Without spare parts in either depots or pipelines, extensive combat flying could only have been maintained by mass cannibalization of existing aircraft—a process that would have led to the slow collapse of fighting capabilities. Hitler's and Göring's failure to recognize the importance of devoting a substantial portion of industrial production to supply reserves caused these inadequacies. Göring had ignored his staff's recommendation that 20 to 30 percent of production be devoted to providing adequate inventories of spare parts and instead, production focused exclusively on building up frontline strength.[29]

In September 1938, the Luftwaffe faced a complicated strategic situation for which it was not prepared. Contrary to what many pro-appeasement historians and contemporary RAF officers supposed, the Luftwaffe's first task was to aid in the destruction of Czechoslovakia. Other tasks such as disrupting French mobilization, protecting Germany's North Sea trade and bombing Britain were strictly peripheral to the central Luftwaffe mission: to aid the army in destroying Czechoslovakia. German planning for both *Fall Rot* and *Fall Grün* deployed the bulk of German aircraft against Czechoslovakia. Only after Czechoslovakia had been destroyed by a decisive air and land attack would the Luftwaffe shift to tasks in the West.[30] This factor must be considered in any evaluation of the Luftwaffe's capability to launch an attack against

Great Britain later in the year.

Bad weather occurred during the projected period of the invasion of Czechoslovakia. For the twelve days from 30 September to 11 October, there were six days of rain and six of fog in the regions surrounding the republic.[31] During the critical first five days of the projected invasion period, the weather was even worse.[32] Young and inexperienced pilots could fly in this weather only with serious losses, especially because so many of them were not yet instrument rated.

Third Air Force, assigned to support the German advance into Czechoslovakia from the southern flank, reported that it would have faced similar obstacles. The phased mobilization of German forces had fully alerted the Czechs to German intentions, and the Czechs had, as a result, deployed their air units from peacetime bases to satellite fields, thus robbing the Luftwaffe of an opportunity to make a surprise attack. Nevertheless, Third Air Force thought that attacks aimed at the Czech ground organization and supply system would eventually have restricted Czech operations, but warned that committing large numbers of aircraft in bad weather would have resulted in unbearable losses through accidents, crashes, and mid-air collisions.[33]

First Air Force, deployed in Saxony and Silesia, reported that while the Czechs were inferior in air power, they could have posed stiff opposition. It had expected to meet a strong, well-organized Czech anti-aircraft defense system over the fortifications and important industrial centers.[34] Thus, although the Luftwaffe would undoubtedly have played a major role in any conquest of Czechoslovakia, because of bad weather and its own inadequacies it would have suffered losses which would have severely crippled its capacity to meet the test of a major European war.

When the Czech crisis began to take on wider implications Germany discovered its unpreparedness to launch a "strategic" bombing offensive. Ironically the Luftwaffe staff had not even begun to plan for such an eventuality.[35] In August 1938, one member of Second Air Force's staff—which would have had responsibility for operations over the North Sea and against Britain—characterized his command's capability as no more than an ability to inflict pinpricks.[36] Another memorandum in September 1938 confessed that there was no possibility of launching air operations against the British Isles which had any prospect of success.[37] As late as May 1939, General Felmy, commander of Second Air Force, complained that the maintenance and supply base in his zone of operations was entirely inadequate to support a major air offensive against Britain. He stressed that should war break out before 1942, Germany would have insufficient forces for such a task, and that in 1939 preparations for such an offensive were "completely inadequate" (*völlig ungenügend*).[38]

In view of aircraft capabilities and aircrew technical skills, Second Air Force declared in August 1938 that Belgium and Holland would have to be seized before there would be any prospect of successful attacks on strategic objectives in Great Britain.[39] This was an accurate forecast, as even in 1940 the only way that the Luftwaffe was able to launch an offensive against Britain was through control of Belgium, Holland, and Northern France. General Felmy warned the Luftwaffe high command in late September that the measures that his command could initiate would merely disturb the English. "Given the means at his disposal a war of destruction against England seemed to be excluded."[40] As late as May 1939, Second Air Force staff was to come to the same conclusion.[41]

In September 1938, the Luftwaffe found itself facing a series of problems which would have severely restricted its effectiveness in any operations against the British Isles. In order to predict weather conditions, the German weather service depended on reports from England, Iceland, Norway and Greenland; ship reports from the Atlantic; and reports from France, Spain and Portugal. At the end of September the Luftwaffe discovered, much to its consternation, that no alternative precautions had been undertaken to replace reports from England, France, ships in the Atlantic, and North America. Thus, it was discovered that once those key reporting stations were lost on the outbreak of war, the weather service would have been operating in the dark—not only forced to base its predictions of British weather on sketchy information, but also hampered even in its ability to predict accurately the weather over Europe.[42]

Even with its well-trained aircrews, the Luftwaffe did not have the capability to bomb accurately in bad weather. Kesselring admitted in 1939 that the "excellent" all-weather aircraft at the disposal of his aircrews would still not enable them to bomb effectively in bad weather.[43] Moreover, in 1938 the Luftwaffe did not possess the radio technology necessary to carry out night or bad weather attacks,[44] while most of the navigational equipment positioned along the North Sea was designed to help aircraft operating over the sea rather than to aid bombing attacks on Great Britain.[45]

Considering that air war has proven to be the most technical and precise form of war man has waged, the technical as well as the operational inadequacies of the Luftwaffe in 1938 define its weaknesses. One cannot help but conclude that the Luftwaffe could not have launched a significant bombing offensive against Britain in 1938. In nearly every respect it was unprepared for such a task and so could not have significantly damaged the British war effort in spite of the often cited weakness of the RAF's Fighter Command. Moreover, the Luftwaffe would even have had difficulty in fulfilling its operational commitments to support German ground forces against Czechoslovakia and in the west.

NOTES

1.   See among others, Keith Eubank, *Munich* (Norman, 1963); Basil Collier, *History of the Second World War* (London, 1965); Keith Robbins, *Munich* (London, 1968); Lawrence Thompson, *The Greatest Treason* (New York, 1968); Sir John Slessor, *The Central Blue* (New York, 1957); Air Marshal Sholto Douglas, *Combat and Command* (New York, 1966).

2.   PRO CAB 24/273, C.P. 283 (37), 29.11.37., p. 141.

3.   Milch Collection, Imperial War Museum, reel 55, vol. 57, chef des Nachschubsamts, Nr. 3365/g.Kdos., 3.11.38.

4.   William R. Emerson, "Operation Pointblank," Harmon Memorial Lectures, No. 4 (Colorado Springs, 1962), p. 41.

5.   Charles Webster and Noble Frankland, *The Strategic Air Offensive Against Germany*, I, *Preparation* (London, 1961), p. 21.

6.   L. E. O. Charlton, C. T. Garrett, and Lt. Cmdr. R. Fletcher, *The Air Defence of Great Britain* (London, 1937), pp. 170-71.

7.   C. Webster and N. Frankland, *The Strategic Air Offensive Against Germany*, I, *Preparation*, p. 79.

8.   Sir John Slessor, *The Central Blue*, p. 223.

9.   Basil Collier, *The Defence of the United Kingdom* (London, 1957), p. 65.

10.   PRO CAB 24/279, C.P. 218 (38), 25.10.38., p. 131.

11.   Air Ministry, Air Historical Branch, Translations, vol. VII, G. 302694/AR/9/51/50.

12.   PRO CAB 3/8, p. 2, 301-A, 14.11.38., CID.

13.   Paul D. Deichmann, *German Air Force Operations in Support of the Army*

(New York, 1968), pp. 9-13.

14.   OKM, B. Nr., 1. Abt. Skl. Ia 961/38 g.Kdos., Berlin, 14.7.38. (USNA, T-1022/2957/PG 48902).

15.   BA/MA, RL 2 II 24, Chef 1. Abt., 22.11.39.

16.   Karlheinz Kens and Heinz J. Nowarra, *Die Deutschen Flugzeuge, 1933-1945* (Munich, 1964), pp. 15, 416.

17.   Richard Suchenwirth, *The Development of the German Air Force, 1919-1939* (New York, 1969), p.97.

18.   BA/MA RL 2 II/115, Luftwaffeführungstab, Az. 89a Nr. 3400/38 g.Kdos., 1 Abt. III, 4, 1.12.38.

19.   Milch Collection, Imperial War Museum, reel 55, vol. 57, 3.6.38., "Verläufiges Flugzeug—Beschaffungs Programm."

20.   Letter from the Reichsminister der Luftfahrt und Oberbefehlshaber der Luftwaffe, 10.8.38., (USNA, T-79/24/000606).

21.   Air Historical Branch, Air Ministry, vol. VII, Translation: Luftwaffe Strength and Serviceability Statistics, G.302694/AR/9/51/50.

22.   Air Ministry, *The Rise and Fall of the German Air Force, 1933-1945*, Air Ministry Pamphlet #248 (London, 1948), pp. 19-20.

23.   BA/MA RL 7/164, Der kommandierende General und Befehlshaber der Luftwaffengruppe 3., 1.12.38.

24.   Ibid.

25.   Air Historical Branch, Air Ministry, vol. VII, Translations: Luftwaffe Strength and Serviceability Statistics.

26.   Milch Collection, Imperial War Museum, reel 55, vol. 57, Der Chef des Nachschubsamts, Nr. 3365/38, g.Kdos., 3.11.38.

27.   Ibid.

28.   Ibid.

29.   Richard Suchenwirth, *The Development of the German Air Force, 1919-1939*, p. 148.

30.   BA/MA RL 7/64 Planstudie 1938, Hauptteil III Aufmarsch und Kampfenweisung "Fall Rot" zu Lw. Gruppenkommando 3., As Plst. 38/Ia op. Nr. 450/38 g.Kdos., 2.6.38.

31.   KTB Arbeitsstab Leeb (USNA, T-79/16/396).

32.   BA/MA RL 7/164, Der kommandierende General und Befehlshaber der Luftwaffengruppe 3., Ia. Nr. 7829/38 g.Kdos., 1.12.38.

33.   Ibid.

34.   BA/MA RL 7/1, Der kommandierende General und Befehlshaber der Luftwaffengruppe 1., Ia. Nr. 197/38, g.Kdos., 11.7.38.

35.   L.W. Gr. Kdo. 2., Führungsabteilung, Nr. 210/38, g.Kdos., 22.9.38.

36.   Vortragsnotiz uber Besprechung mit Ia des Befehlshabers der Luftwaffengruppe Braunschweig 25.8.38. (USNA T-1022/2307/34562).

37.   L.W. Gr. Kdo. 2., Führungsabteilung, Nr. 210/38, g.Kdos., 22.9.38.

38.   BA/MA, RL 7/42, Luftflottenkommando 2., Führungsabteilung, Nr. 7093/39, g.Kdos., Braunschweig, 13.5.39., pp. 4, 22.

39.   Karl Gundelach, "Gedanken über die Führung eines Luftkrieges gegen England bei der Luftflotte 2. in den Jahren 1938- 1939," *Wehrwissenschaftliche Rundschau* (January 1960).

40.   L.W. Gr. Kdo. 2., Führungsabteilung, Nr. 210/38, g.Kdos., 22.9.38.

41.   Ob.d.1, Generalstab, L. Abt., Nr. 5095/39, g.Kdos., 22.5.39.

42.   See particularly BA/MA RL 7/50, IW/38, g.Kdos.; IW 23/29 G. Vortrags Planspiels 10-14.5.39.; BA/MA RL II/101 Luftwaffengruppenkommando 2., 6.12.38., Führung Abt/IW B.Nr. 129/38, g.Kdos.

43.   BA/MA RL 2 II/101, Vortrag: General der Flieger Kesselring, 1.3.39.

44.   Air Historical Branch, Air Ministry, AHB 6 No. VII/153, "German Air Force Policy during the Second World War," a review by Oberst Bernd von Brauchitsch, p. 3.

45.   Karl Gundelach, "Gedanken über die Führung eines Luftkrieges gegen England bei der Luftflotte 2. in den Jahren 1938/1939," p. 35.

CHAPTER FOUR

THE AIR DEFENSE OF GERMANY:
DOCTRINE AND THE DEFEAT
OF THE LUFTWAFFE

Historians have generally viewed the conduct of Germany's air defense in World War II as an ad hoc, reactive response to the swelling Allied air offensive of 1943 and 1944. There is much to support their point of view. Nevertheless, evidence suggests that the influence of prewar doctrine and thinking on air power played an essential element in the German response to the Combined Bomber Offensive. Consequently, to understand fully the German strategic response to Allied efforts to wreck the Reich's cities and economy requires a clear understanding of the mental framework and attitudes toward air power of Germany's political and military leaders. Otherwise that response seems both senseless and bizarre.

Until recently, historians have pictured German air doctrine as oriented almost exclusively toward ground support, but in fact, like other air forces in the 1930s, the Luftwaffe was thinking in terms of "strategic" bombing.[1] Unlike the RAF and the U.S. Army Air Forces, however, Germany's exposed continental position forced the Luftwaffe to address a wider doctrinal framework. Thus, the Germans paid significant attention to such "tactical" air missions as close air support, interdiction, and air superiority.

Nazi ideology and its manifestation in an aggressive foreign policy also contributed to the German view of air war before the Second World War. In particular the Germans felt that Nazi ideology with its emphasis on a Volksgemeinschaft, provided a national psychological strength that would enable them to better endure aerial bombardment than their future opponents in Europe. In May 1933, Dr. Robert Knauss, one of Erhard Milch's chief subordinates in Lufthansa, forwarded an important memorandum arguing for the creation of a "strategic" bomber force to his superior. Knauss suggested that "the terrorizing of the enemy's chief cities and industrial regions through bombing would lead that much more quickly to a collapse of morale, the weaker the national character of his people is, and the more that social and political rifts cleave his society." He then argued that the totalitarian society of Nazi Germany would prove far more capable of absorbing and enduring "strategic" bombing attacks than would the fractured societies of Great Britain and France.[2]

Nazi Germany's aggressive foreign and strategic policies were obvious from the earliest days of the regime.[3] For the Luftwaffe this led to a rapid buildup and a more expansive theory of airpower employment than was the case either in Britain or the United States. The first chief of staff, General Walther Wever, while a strong supporter of the creation of a "strategic" bombing capability, seems to have been more willing to recognize the limitations that combat conditions would impose on such a use of airpower. In fact, the Luftwaffe's doctrinal manual, *"Die Luftkriegführung,"* suggested that "strategic" bombing attacks might take too long to render effective military support to the overall military situation.[4]

In terms of Germany's overall strategic situation, the Luftwaffe aimed to support the forward thrust of the Reich's military forces. Its first operational objective would be the gaining of air superiority over the theater of operations by attacks on the enemy's base structure and airfields and by an aggressive air-to-air campaign against his air forces. Once air superiority had been won, the Luftwaffe would turn to attacking enemy industry and population centers, interdiction missions, and if the army needed help, close air support strikes. The German conception of air war, at least in Central European terms, envisioned an interdependent campaign in which ground forces would play a major role in striking at enemy centers of power, including enemy air power.

Where did air defense fit into the German scheme of air war? Here, the Germans showed considerably more realism conceptually than did most airmen in Britain or the United States in the prewar period.[5] Within a year of Hitler's coming to power, a war game had shown Germany's strategic vulnerability to air attack. The exercise suggested that a German bomber fleet could not immediately destroy an enemy's air force. Consequently, in any major war, Germany's cities and industry would be vulnerable to air attack. Unlike the RAF's air staff, who throughout the 1920s and early 1930s argued that air defense measures would be a waste of effort,[6] the Luftwaffe officers who analyzed the 1933-34 war game concluded that strong fighter forces, as well as anti-aircraft guns would be necessary to protect the Reich.[7]

By the outbreak of the war the Luftwaffe had evolved into the largest and most impressive air force in the world. Its force structure and approach to air war were aggressive and wedded to the offensive. Nevertheless, for a variety of reasons it was very much an air force constrained in its mental framework to a conflict within the confines of central Europe. Its "strategic" bombing capabilities were clearly the most advanced in the world in terms of capability and in some respects in terms of its technology.[8] However, the Luftwaffe's follow-on "strategic" bomber, the four-engine He 177, turned into a technological nightmare and dead end.

Germany's approach to air defense, however, was considerably different from that of Great Britain. To begin with the Germans had more faith in the anti-aircraft gun than did the British to meet the possibility of enemy air attacks on German industry and population centers.[9] Radar had been developed in Germany in the late 1930s, but the German air defense system was not equal to Fighter Command's system for the defense of the United Kingdom. The obvious reason was that Germany planned to be the attacker in the coming war, and with the exception of Britain, their continental enemies were vulnerable to German land as well as air assault. For the defense of major industrial and population centers, the Luftwaffe planned to rely on its large concentrations of anti-aircraft guns. Here the lessons of the Spanish Civil War proved misleading. With the generally good weather and the relatively low level at which bomber formations had attacked targets, the Luftwaffe overestimated the performance of anti-aircraft weapons, thereby distorting future programs for the air defense of the Reich.[10]

The one area where the Germans had developed a real radar- guided air defense system lay in the Heligoland Bight. That system provided protection for the navy's fleet units and major bases in the North Sea and within this limited context the system proved effective. In December 1939 the RAF launched a major raid of Wellington bombers across the North Sea in the hope of catching major units of the German navy. The constraints under which the raid was launched—the attacking bombers could not attack any German navy ship tied up to land for fear that civilians might be injured or killed—provided little prospect for success.[11] The German air defenses then ensured that this hopeless mission suffered extraordinarily high casualties. Bf 109s guided by German radar stations picked up the incoming Wellingtons and in perfect daylight conditions made the interception. Ten Wellingtons were shot down; nearly all of the remainder were damaged. The bombers did sight one battleship, one heavy cruiser, one light cruiser, and five destroyers tied up at docks. The Wellingtons took photographs, but none dropped their bomb loads because, "as the ships were lying in close proximity to buildings, no attempt was made to bomb them."[12]

The salutary experience of this unescorted bomber mission along with other experiences early in the war went a long way toward convincing the RAF that unescorted, daylight, deep penetration bombing attacks simply were not in the cards. Consequently, when the new Churchill government based much of its post-17 June 1940 strategic approach on a "strategic" bombing offensive against Germany, the RAF embarked on this effort at night. Ironically, Bomber Command had done so little preparation for night bombing that it took over a full year for the British to discover that these efforts were almost completely ineffective. In August 1941 an analysis of mission photographs suggested that only one in three British aircraft was hitting within seventy-five square miles of the target.[13] In fact, given the almost complete lack of navigational aids, Bomber Command had a difficult time in hitting even cities. On 1 October 1941 with Karlsruhe and Stuttgart as targets, British bombers "were reported over Aachen, Eupen, Malmedy, Koblenz, Neuweid, Kreuznach, Frankfurt am Main, Wiesbaden, Limburg, Darmstadt, Mainz, Trier, Offenburg," and thirteen other cities.[14]

As the British offensive floundered in 1941 and for most of 1942, the Germans evolved a somewhat effective air defense system that imposed a rising rate of attrition on Bomber Command. In July 1940, partially in response to British raids on Germany, the Luftwaffe created the 1st Night Fighter Division in Brussels under General Joseph Kammhuber.[15] This German defensive force, like the night fighter force across the

Channel, represented an ad hoc conglomeration of Bf 109s, Do 17s, and a combined flak-searchlight regiment. The initial conception was that the bombers would attack Bomber Command's bases in Britain, while the flak-searchlight regiment would illuminate British bombers for the fighters and the anti-aircraft guns. Of its component parts, the second was less than satisfactory. German night fighters and searchlights were about as successful in finding British bombers as British bombers were in locating German cities. However, the intruder missions flown against Bomber Command bases showed great promise through the summer of 1941. At that point alarmed by bomber losses in the Russian campaign, Hitler ordered the intruder missions halted and for the remainder of the war rarely allowed attacks on British or American bomber bases. Ironically, in view of the historiographical view that the Germans possessed no notion of "strategic" air bombing, Hitler's response fits neatly within the Douhetian and Trenchardian prewar definitions of air strategy. In other words, from the beginning Hitler's attitude toward the enemy's "strategic" bombing denigrated efforts aimed at attacking the force itself.[16] Such an approach that rejected what would today be characterized as a "counter force" strategy remained a hallmark of the German response to "strategic" bombing attacks for the remainder of the war.

Beginning in October 1940, the Germans introduced Würzburg radar units into the struggle by placing the first set in Holland. By late 1941, Kammhuber had established a belt of radar stations stretching from Denmark to Holland and then south through Belgium into northern France. Like the British night defense system, the German defenses consisted of a tightly controlled marriage between a GCI (ground controlled intercept) station and one fighter operating in the same designated area that formed a portion of the larger belt. Through 1941 the system developed into a formidable threat to Bomber Command's operations. This approach did have one obvious weakness. With only one German GCI station and fighter over a given area, Bomber Command was in a position to swamp the defenses if it could feed its aircraft through the GCI boxes in a concentrated stream.

For most of 1942 it was not able to do so, nor was it able to find and hit city targets on a consistently effective basis. There were, however, successes that should have troubled the German high command. The Lubeck and Rostock raids of spring 1942 did significant damage to those urban areas. Even more successful was Air Marshal Arthur Harris's "Thousand Bomber" raid on Cologne in May 1942, which destroyed much of the downtown area. The Luftwaffe exacerbated Hitler's anger at the devastation on the ground by suggesting that a victory bulletin be issued because it had downed thirty-seven bombers, for "a 50 percent success." The Führer pointedly noted that not just the RAF's claims of a 1,000 bomber raid, but the damage on the ground suggested that the attacking force was enormously larger than Luftwaffe estimates.[17] On 3 June, Hitler received the Luftwaffe's Chief of Staff, Hans Jeschonneck, and gave him a severe dressing down. Interestingly, the Führer suggested that the only reply to such "terror" raids was retaliation in kind.[18]

In some ways, it was fortuitous that Bomber Command was not able to repeat the success of the 1,000 aircraft attack on Cologne for the remainder of 1942. As the British raids over the rest of the year achieved little more than they had in 1941, the full import of Bomber Command's threat slipped from the German consciousness. Kammhuber and his defense system received no greater allocation of resources than before.

Instead of beefing up the night defense forces, the Germans responded to the spring 1942 raids as they would for much of the war: by launching a series of night retaliatory raids against British cities. These raids, derisively termed "Baedecker" raids by the British, achieved little. The damage that occurred was achieved at the expense of heavy Luftwaffe aircraft and crew losses, especially in the training units, which lost many irreplaceable instructional crews.[19]

In 1942, the Reich had escaped relatively undamaged from the much heralded allied "strategic" bombing offensive. But in 1943 the roof began to fall in by day as well as by night. For clarity, we will continue our analysis of the German response to night bombing through summer 1944 before turning to the response to the wholly different set of problems raised by the daylight precision bombing attacks of the U. S. Army Air Forces.

The forces that Bomber Command launched against Germany's cities in 1943 represented an enormous jump technologically, operationally, and quantitatively over those of earlier years. Beginning in March 1943 and for the next three months the RAF battered the Ruhr with "a whole series of consistent and pulverizing blows among which the failures were much rarer than the successes."[20] Nevertheless, German night defenses kept pace, but barely. Despite its successes, Bomber Command was on the thin edge of unacceptable losses. In forty-three major attacks over the spring during the Battle of the Ruhr, the RAF lost no less than 872 bombers with 2,126 damaged.[21] This high rate of loss led the British to introduce "Window" (code name for chaff) in the massive series of raids launched against Hamburg at the end of July 1943.[22] With perfect conditions for the raid of 27 July, Bomber Command came close to annihilating the German port city. Between 30,000 and 40,000 people perished;[23] and Albert Speer was led to exclaim to Hitler that six more attacks on such a scale would "bring Germany's armaments production to a halt."[24]

On the operational level, the Luftwaffe adapted with considerable speed to the disruption of its defenses. Even before Hamburg, some within the Luftwaffe had been pushing for a more flexible approach than the GCI-based Kammhuber line. Major Hajo Hermann suggested the use of a concentration of day fighters along with the searchlight force over the cities being attacked as a possible means of increasing British losses.[25] Others had argued for a more fundamental restructuring of German defenses. Göring's staff was urging better radar sets to support a night fighter force that would not be directly tied to GCI sites but which would search out and follow the bomber stream.[26]

The British success over Hamburg concentrated German minds for the short term. This concentration was undoubtedly helped by the swelling daylight offensive of the U.S. Army Air Force. The Luftwaffe's chief of production, Field Marshal Erhard Milch, warned his staff in the air ministry on 30 July that Germany could only look forward to an intensification of the enemy's air offensive. Germany must master the threat or face a desperate situation. Milch further announced that Hitler had put top priority on air defense and on production of day and night fighters as well as on Flak.

The problem was that such an emphasis rapidly broke down in the face of conflicting pressures and the already established proclivities of Germany's military and political leadership. Even Milch, who would remain the most clearheaded German leader in evaluating the danger throughout the period, was calling loudly in March 1943 for reprisal raids against Great Britain. As he told his staff, "Our entire armaments effort . . . is dependent on whether we can clear our own skies by carrying out the appropriate attacks on the British home base—either on their airfields or on their industry or on their civilians and cities."[27]

Hitler's thinking throughout the remainder of the war remained fixed on the Douhetian notion that retaliation was the only effective means of ending bomber attacks. Shortly after the Hamburg catastrophe the Führer warned his military aides: "Terror can only be broken with terror." Attacks on German airfields made no impression on him, he suggested, but the smashing of the Reich's cities was another matter. It was the same thing with the enemy. "The German people demanded reprisals."[28]

Mixed in with Hitler's attitude toward air defense was a consistent unwillingness to recognize the threat. In July 1943, Kammhuber had presented the Führer with a proposal for a radical restructuring of Germany's air defenses to meet massive Allied air production. Hitler, however, demanded to know the origin of these "crazy numbers" which, he assured Kammhuber, were false.[29]

This emphasis on retaliation and unwillingness to give air defense the priority that it demanded led the Germans to make the wrong choice. In summer 1943 they could have radically restructured their aircraft industry for a massive output of day and night fighters at the expense of other aircraft types. The other choice, the continued emphasis on retaliation weapons, meant the loss of the air war in the skies over Germany. Milch himself had suggested a target of 5,000 fighters per month to Hitler in March 1943.[30] But Hitler and Göring were unwilling to address the "strategic" bombing threat with a defensive response. Rather the Germans continued to maintain the same level of bomber production, and those bombers would be used up in pointless pinprick "Baby Blitz" raids on London in early 1944.

Even greater resources were poured at this time into the revenge weapon programs, the so-called V-1 and V-2 (*Verwaltungswaffe* [revenge weapon]-1 and 2). As Albert Speer suggested to an enthusiastically appreciative audience in May 1943 in the Ruhr: while "German mills of retribution may often seem to grind too slowly, they do grind very fine. . . ." Speer had just seen a successful firing of what would become the V-2, and his continued support for the rocket program throughout 1943 and 1944 caused a major diversion of production capacity and raw materials that would have been far better spent in defending German air space.[31]

The factors that led the German high command to search for a Douhetian solution to the air defense problem were not only their leadership's peculiarities and attitudes, but also the reaction of the German population to the bombing. Recent scholarship in the Federal Republic indicates that as early as summer 1942, night bombing was having a serious impact on German morale.[32] In 1943, the *SD* (*Sicherheitsdienst*,

Secret Police) reports on what the population was saying (reports which were widely read in the highest levels of Nazi leadership) noted that the population was restive, angry, and above all demoralized by the attacks. Among the many manifestations of their demoralization were the comments of women "of the lower classes" that even 1918 had not been as bad as this.

In one of those quirks of human nature, at the same time that the German population was depressed and gloomy over the bombing, it was also extraordinarily angry at its tormentors. Throughout the last years of the war the German people were demanding retaliation against Britain for the damage inflicted to their houses and cities. That anger and demand for retaliation helps to explain the Nazi response and the fact that there was no massive infusion of German resources into the air defense system. In October 1943, Göring suggested to his staff that the German population did not care whether the Luftwaffe attacked British airfields. "All they wished to hear when a hospital or a children's home in Germany is destroyed is that we have destroyed the same in England; then they are satisfied."[33] In October 1943 Göring reproached Milch for placing too much emphasis on the Reich's air defense and for robbing production from the bomber force.[34] On 23 November, the *Reichsmarschall* allocated most of the Ju 388 production to the bomber units rather than to the night fighter force.[35] By the end of that month, he deliberately held down the production of fighters in favor of bombers. As he commented to Fritz Sauckel, slave labor procurer for the Reich, the Luftwaffe had to have bombers:

> **Göring:** I cannot remain on the defensive; we must also have an offensive. That is the most decisive.

> **Sauckel:** The only argument that makes an impression on a racial cousin [the British] is that of retaliation.[36]

If the resources devoted to bomber production subtracted from what the Germans might have been able to devote to air defense, the V-1 and V-2 weapons represented an even greater misallocation of resources. The strategic bombing survey estimated that the industrial effort and resources expended for these weapons alone in the last eighteen months of the war equaled the production of 24,000 fighter aircraft.[37] Here the regime was reacting to popular pressures, and the resulting decisions responded to political factors and the *Weltanschauung* of the leadership rather than to the strategic and military realities that confronted Germany.

Interestingly, the Germans also maintained a heavy emphasis on the anti-aircraft artillery defenses for political reasons similar to those that kept the emphasis on a "retaliatory" response. By 1942, it was clear that anti-aircraft guns against high-altitude bombers could not achieve the same level of success that they had in the Spanish Civil War. Nevertheless, because of the reassurance that massive anti-aircraft barrages provided to the civilian population, the Germans continued to expand their anti-aircraft forces throughout the war.[38] Despite Luftwaffe objections, Hitler was demanding a drastic strengthening of German Flak forces as Bomber Command laid waste the Ruhr in its spring 1943 offensive.[39] At the same time Goebbels, with the support of the Gauleiters (Nazi party district leaders), was berating Milch because of the supposed numerical insufficiency of anti-aircraft guns to defend Germany's cities.[40] The result of such pressure led to an increase in the Reich's anti-aircraft artillery forces from 791 heavy batteries in 1940, to 2,132 by 1943—a huge investment in equipment and material.[41] All of these batteries expended vast amounts of ammunition with results more visually spectacular and alarming to air crews than effective in their results. It took the 88 mm flak 36 weapon somewhere over 16,000 shells to bring down one aircraft flying at high altitude.[42]

The German reaction to the American strategic bombing offensive reflected much of the same attitudes that Bomber Command's nighttime offensive brought forth. In this case, however, the American "strategic" bombing campaign reflected a more direct, straight out attack on the Luftwaffe. To use modern terms, its strategic approach was more of a "counterforce" strategy while Bomber Command's efforts represented very much a "countervalue" strategy. The Americans had, in fact, made little secret of what they were planning to do to the Germans. Certainly, American propaganda made clear American production goals for bombers in 1943. The German leadership, however, in public and internal utterances casually and blatantly dismissed such numbers as fantasy.[43] Göring suggested that the Americans "could only produce cars and refrigerators."[44]

Consequently, the full weight of the American bombing offensive broke on the Germans in the summer of 1943 almost completely unexpectedly. In January 1943, Reichsmarschall Göring suggested an increase in the day fighter force, not because of worries over Allied aircraft production, but rather to bolster fighter bomber support for the army on the Eastern Front.[45] Even Adolf Galland, who was pushing for a major increase in the fighter force, did not fully recognize the danger in the west. At the same time he predicted that the main weight of the air war would remain in the Mediterranean throughout the year.[46] As a result of such attitudes the Luftwaffe made a massive commitment of the fighter force to the defense of Tunisia, Sicily, and Italy over the first half of 1943. Why Galland was articulating such a point of view is clear from a remark that he made in February that suggested that his units in the west had solved the problems posed by the four-engine daylight bomber offensive.[47] Unfortunately for the Germans, Galland had only experienced limited American probes in 1942; the American offensive in the summer of 1943 would reflect the first fruits of the massive American production.

In view of the scarcity of German resources throughout Europe the commitment of the Luftwaffe to the Mediterranean reflected a gross overcommitment. Equally sizable fighter forces were committed to air operations on the eastern front. Not only did the Luftwaffe suffer severe losses in supporting army operations in the Stalingrad/Kharkov battles of winter 1943 and the Kursk offensive of summer 1943, but it embarked in spring 1943 on a costly air superiority battle over the Kuban Peninsula that served no discernable purpose.[48] The massive losses suffered in these battles on the periphery exhausted the Luftwaffe day air superiority fighter force just at that moment when the American "strategic" bomber force was emerging as a direct and dangerous threat to German war production itself.

The immediate threat in 1943 was mastered by desperate halfmeasures that saw the Luftwaffe almost entirely shut down its air effort in the Mediterranean and on the Eastern Front. Almost from the first the Germans showed a proclivity for throwing their night fighter force, despite its expensive and sophisticated aircraft and specialized aircrews, into the daylight air battle. As early as April 1943, an Ultra intercept indicated a willingness to use the night fighters in the daylight air battles. Göring kept only night fighter pilots with more than twenty victories from participating in daytime operations.[49] In the Anglo-American air attacks of 17 and 18 August, the Luftwaffe lost thirty night fighters with thirty-five more damaged in day and night operations. Twenty-one of those aircraft were lost in daylight operations alone, and a senior staff officer remarked that the Bf 110 should not be used in the daytime when it might come in contact with British or American fighters--a remark that might have had some uniqueness in the summer of 1940 but seems somewhat out of place in 1943.[50]

In early 1943, the Luftwaffe, with major commitments in the Mediterranean and on the eastern front, had left the day defense of the west on the same basis on which it had rested in previous years. Some 250 to 300 fighters, scattered from Holland to Brittany, scrambled in small formations to meet the American thrusts.[51] The resulting lack of fighter concentration made it difficult to dent the Fortress formations and put the Germans at considerable disadvantage in fending off Allied fighters. By June, the Luftwaffe's western fighter forces were breaking down as *Luftflotte* 3 (Third Air Force) reported that its fighters were suffering heavy losses in intercepting Fortress formations accompanied by numerous fighters.[52] During that month, after a trip to the west, Milch reported that morale among the fighter pilots was excellent, but that the figure for the number of fighters available was "much too low." He urged that the Luftwaffe quadruple fighter forces in the west and that at a minimum a full month's production of Bf 109s and FW 190s go to units in western Europe.[53]

Here again, as with the response to the night offensive, the Luftwaffe responded with too little, too late. By concentrating almost all of its day fighter force and by throwing much of the night fighter force into the battle in a haphazard fashion it was able to stave off defeat in fall 1943 and deal Eighth Air Force some significant body blows. It was not, however, able to overcome the threat. It was able to blunt only the deep penetration, unescorted raids, and throughout summer and fall 1943 the range of American fighter escorts crept ever deeper into the Reich. The German leadership, particularly Göring, simply refused to accept the growing depth of Allied fighter penetration.[54] To have done so would have forced the Luftwaffe leadership to accept the argument for a massive concentration of resources on a day fighter air superiority strategy and the abandonment of the Douhetian retaliatory strategy. Hitler and Göring were unwilling to make such a decision. The result in terms of the great winter and spring 1944 air battles was a foregone conclusion.[55] Those battles and the relentless pressure that the daylight forces of the USAAF and the RAF were able to place on the Luftwaffe led to the complete collapse of the German air defense system by June 1944.

The failure of the German air defense system in the Second World War must be attributed to a general failure of the German leadership to fight that war in classical terms. In 1924, the British air staff argued that air forces employed in attacking an enemy nation

> can either bomb military objectives in populated areas from the beginning of the war, with the objective of obtaining a decision by moral[e] effect which such attacks will produce, and by the serious dislocation of the normal life of the country, or, alternatively, they can be used in the first instance to attack enemy aerodromes with a view to gaining some measure of air superiority and, when this has been gained, can be changed over to the direct attack on the nation. The latter alternative is the method which the lessons of military history seem to recommend, but the Air Staff are convinced that the former is the correct one.[56]

This countervalue strategy very much characterized the air defense response of the Reich to the problems occasioned by the Combined Bomber Offensive. Ironically in view of the current historical view of the Luftwaffe, its defeat in World War II lay consequently in the German leadership's wholehearted acceptance of the Douhetian, Trenchardian view of air war rather than in an antiquated "pro-army" conception of air doctrine and air power. By refusing to meet the "strategic" bombing offensives of its enemies with a military counterforce strategy, the Germans insured their defeat in the skies over Europe in early 1944 and the almost total destruction of their cities and war production.

NOTES

1.   For the basic groundbreaking work on this point I am indebted to a lecture given in September 1980 at the Air War College, Maxwell Air Force Base,

Alabama, by Oberstleutnant Klaus Maier of the Militärgeschichtliches Forschungsamt, Federal Republic of Germany. For a detailed examination of Maier's arguments see Klaus A. Maier, Horst Rohde, Bernd Stegemann, and Hans Umbreit, *Das Deutsche Reich und der Zweite Weltkreig*, Vol. II, *Die Errichtung der Hegemonie auf dem europäischen Kontinent* (Stuttgart, 1979), pp. 43-70. See also my article, "The Luftwaffe Before the Second World War: A Mission, A Strategy?," *Journal of Strategic Studies* (September 1981).

2.   Bernard Heimann and Joachim Schunke, "Eine geheime Denkschrift zur Luftkriegskonzeption Hitler-Deutschlands vom Mai 1933," *Zeitschrift für Militärgeschichte*, vol. III (1964), pp. 72-86.

3.   "Aufzeichnung Liebmann," *Vierteljahrshefte für Zeitgeschichte*, 2, no. 5, (October 1954).

4.   "Die Luftkriegführung," Berlin 1935; copy made available to the author by Oberstleutnant Klaus Maier of the Militärgeschichtliches Forschungsamt.

5.   On the air defense question, the RAF's Sir Hugh Dowding proved himself to be enormously farsighted in the technological development of Fighter Command. However, Dowding was an exception in the RAF and only the wholehearted support of the Chamberlain government for an air defense force in the 1937-39 period forced the RAF to place significant emphasis on the creation of an air defense force.

6.   See particularly Public Record Office, Air 20/40, Air Staff Memorandum No. 11A, March 1924, for a statement of the RAF's dominant attitude towards air defense until the Chamberlain government forced through a substantial buildup of Britain's air defenses.

7.   Karl-Heinz Völker, *Documente und Documentärfotos zur Geschichte der deutschen Luftwaffe* (Stuttgart, 1968), Doc. 184, p. 429.

8.   At the outbreak of the war the Germans were far ahead of the RAF and the U.S. Army Air Force in the developing of blind bombing devices (in particular "Knickebein"). We and the British would not possess similar technology until 1943.

9.   In fairness to the Germans, the British also placed a heavy reliance on anti-aircraft guns throughout the war.

10.   Auswirkung der Erfahrungen im Spanien. "Aus einer Ausarbeitung von Generallt. Galland über die Luftverteidigung des Reiches, 1946," Karlsruhe Collection, Albert F. Simpson Historical Research Center (AFSHRC), K 113.302.

11.   This attitude of the Chamberlain government reflected the whole British approach to the war at this time despite the savaging that the Luftwaffe had given to Warsaw and other Polish cities.

12.   PRO CAB 66/4 WP(40) 1, 1.1.40, "Air Operations and Intelligence."

13.   Sir Charles Webster and Noble Frankland, *The Strategic Air Offensive Against Germany*, vol. IV (London, 1961), Appendix 13, Report by Mr. Butt to Bomber Command on his examination of night photographs, 18.8.41.

14.   Ibid., vol. I, p. 185.

15.   This discussion of the origin of German night fighter tactics is drawn from the extensive review of night air defense written for the RAF by General Joseph Kammhuber at the end of the war, hereafter referred to as "Development of Night Fighting."

16.   See Williamson Murray, *Luftwaffe* (Baltimore, 1985), appendix 1.

17.   For the German reaction to the Cologne raids see *Kriegstagebuch des Oberkommandos der Wehrmacht*, vol. II, ed. Andreas Hillgruber (München, 1982), entries for 31.5., 2.6., and 3.6.42. pp. 394, 398-400.

18.   Ibid. See also Nicolaus von Below, *Als Hitlers Adjutant, 1937-1945* (Mainz, 1980), pp. 311-12.

19.   Air Ministry, *The Rise and Fall of the German Air Force* (London, 1948), p. 196.

20.   Webster and Frankland, *The Strategic Air Offensive Against Germany*, vol. II, p. 108.

21.   Ibid., pp. 110-11.

22.   For the most vivid account of the destruction of Hamburg see Martin Middlebrook, *The Battle of Hamburg, Allied Bomber Forces against a German City in 1943* (London, 1980).

23.   See among others, Hans Rumpf, *The Bombing of Germany* (London, 1963), pp. 82-83; Middlebrook, *The Battle of Hamburg*, p. 272; OKW Wehrwirtschaftsstab, "Erfahrungen bei Luftangriffen," von Oberst Luther, 15.1. 1944, National Archives and Records Services T-79/81/000641.

24.   Albert Speer, *Inside the Third Reich* (New York, 1970), p. 284.

25.   Tag und Nachtjagd, Besprechungsnotiz No. 63/43 am 27.6.43, Obersalzberg, AFSHRC: K113.312.2, v.3.

26.   BA/MA, RL3/54, Der Reichsminister der Luftfahrt und der Oberbefehlshaber der Luftwaffe, Br. 21 Nr 8731/43, 21.7.43, Betr.: "Sofortmassnahmen für Verfolgungsnachtjagd über grössere Räume."

27.   David Irving, *The Rise and Fall of the Luftwaffe* (Boston, 1973), p. 208.

28.   "Hitler zur Frage der Gegenmassnahmen zur Beantwortung der allierten Luftangriffe." 25.7.43, AFSHRC: K113.312, v.3.

29.   Friedhelm Golücke, *Schweinfurt und der strategische Luftkrieg, 1943* (Paderborn, 1980), p. 115.

30.   Irving, *The Rise and Fall of the Luftwaffe*, p. 202.

31.   David Irving, *The Mare's Nest* (Boston, 1965), pp. 58-59, 87-90.

32.   The following discussion is based on an interesting section of a major work on the establishment of the Hitler myth in Bavaria. The author based this study on the SD reports on the attitudes of the population; and as those reports were widely read in the upper levels of the Nazi government, certain aspects of German air defense strategy in the last years of the war became clear. See Ian Kershaw, *Volksmeinung und Propaganda im Dritten Reich* (Stuttgart, 1980), pp. 176-86.

33.   "Heimatverteidigungsprogramm   1943,   Besprechung   beim Reichsmarschall am 7.10.43, Obersalzberg, Fortsetzung," AFSHRC: K113,312-2, V.3.

34.   BA/MA, RL3/61, "Stenographische Niederschrift der Besprechung des Reichsmarschalls mit GL und Industrialrat am 14.10.43."

35.   BA/MA, RL3/61, "Stenographische Niederschrift der Besprechung beim Reichsmarschall am 23.11.43 in Karinhall."

36.   BA/MA, RL3/61, "Stenographische Niederschrift der Bespreschung beim Reichsmarschall an 23.11.43 in Karinhall."

37.   U.S. Strategic Bombing Survey, "V-Weapons (Crossbow) Campaign," Report 60 (January, 1947).

38.   See Murray, *Luftwaffe*, p. 182.

39.   Nicholas von Below, *Als Hitlers Adjutant* (Mainz, 1980), p. 208.

40.   Joseph Goebbels, *The Goebbels Diaries*, ed. Louis Lochner (New York, 1948), entry for 10.4.43, p. 322.

41.   Golücke, *Schweinfurt und der strategische Luftkrieg, 1943*, p. 153

42.   Ibid., p. 156.

43.   See Goebbels, *The Goebbels Diaries*, pp. 41, 65, 104, 169, 251.

44.   Asher Lee, *Goering, Air Leader* (New York, 1972), p. 58.

45.   BA/MA, RL 3/60, Besprechungsnotiz Nr. 8/43, 25.1.43.

46.   Air Ministry, *The Rise and Fall of the German Air Force*, p. 219.

47.   Oberst Vorwald, Chef des Technischen Amtes, 19.2.43, "Kurzer Bericht über die Besprechung beim Herrn Reichsmarschall mit den Flottenchefs vom 15 bis 17. February 1943."

48.   See Hermann Plocker, *The German Air Force Versus Russia, 1943* (Maxwell AFB, Ala., 1967), chapter 2.

49.   *Ultra, History of U.S. Strategic Air Force Europe versus German Air Force* (Frederick, Md., 1980), p. 29.

50.   Auszug aud der G.L.-Besprechung am 20.8.43 im RLM, "Der erste grosse Einsatz der 'wilden Sau'," AFSCHRC: K113.312-2, v.3.

51.   Air Ministry, *The Rise and Fall of the German Air Force*, p. 287.

52.   BA/MA, RL 7/112, Luftflottenkommando 3, Führungsabteilung (I) Nr.

8480/43, 12.7.43., "Tätigkeitsbericht der Luftflotte 3 für den Monat Juni 1943."

53.   BA/MA, RL 3/50, Der Staatssekretr der Luftfahrt und Generalinspektur der Luftwaffe, Gst, Nr. 847/43, 29.6.43, "Bericht über Besichtigungsreise 7.6. bis 12.6.43."

54.   See the exchange between Galland and Göring in Speer, *Inside the Third Reich*, p. 290.

55.   For a fuller discussion of the air battles of 1944 that won complete air superiority for Allied air forces over the European continent, see Murray, *Luftwaffe*, pp. 216-32.

56.   Public Record Office, Air 20/40, Air Staff Memorandum No. 11A, March 1924.

CHAPTER
FIVE

A TALE OF TWO DOCTRINES:
THE LUFTWAFFE'S
"CONDUCT OF THE AIR WAR"
AND THE USAF'S MANUAL 1-1

One of the critical factors facing modern military organizations is the articulation of doctrine: the conceptual framework within which they plan and train their forces in peace and in war for maximum effectiveness in battle. Doctrine is particularly important in giving commanders and subordinates on the battlefield shared assumptions that enable them to know intuitively what others might be doing under the confused pressures of combat.

There have been two general approaches to doctrine in twentieth-century military organizations: at one end of the spectrum the Germans framed their approach to war as broadly as possible. Their doctrinal manuals attempted to suggest, to guide, and to set broadly based criteria for action: they did not attempt to spell out every contingency and expected reaction. The German outlook might best be summed up by what they call *Auftragstaktik* (which can best be translated as mission-oriented tactics).[1] German commanders set missions for subordinates but did not specify how they were to achieve those goals. This approach promoted a decentralized command and control system in which the organization expected everyone from the rank of junior officer and NCO on up the chain of command to act independently within the framework of a general plan of operations. Such a system *demanded* an extremely high level of competence, training, knowledge of weapons systems, and leadership.[2] Without the constant training that exposed officers and NCOs to a wide variety of possible battlefield conditions, the German system would quickly have degenerated into chaos. Constant training aimed to expose German officers and NCOs to as many different combat situations as possible to encourage flexibility and the use of independent judgment.

The American approach to doctrine differed significantly from the German. To begin with, the American mobilizations for World War I and World War II came at the last possible moment and called for an enormous expansion of the officer corps. Not only was the overall quality of the new American officers less than what the Germans would have accepted,[3] but the training time available before American units were committed to battle was not sufficient to give American officers the same level of experience as that of their German counterparts. Thus, in response to the demands of hurried mobilization, the American approach to doctrine has been more rigid and didactic than was the case with the Germans. But beyond these mobilization problems the American military have had a love affair with SOPs (Standard Operating Procedures).

A closer look at American air doctrine during the Second World War suggests that this is so.[4] In the 1930s the instructors at the Air Corps Tactical School articulated a doctrine of strategic bombing in which they argued that large, well-armed four-engine bombers, flying in close, mutually supporting formations would be able to penetrate in daytime deep into enemy territory. On the way they would be able to fight off enemy fighters and suffer acceptable losses. Their targets would be carefully selected as those keystone industries on which the enemy's economic structure depended. The doctrine rested on a large number of assumptions—the relationship of which turned out to be exponential rather than arithmetic.

Because there was so little experience on which to draw, one should not necessarily fault those early thinkers for their lack of realism. Unfortunately, much of the historical criticism has turned on the failure of American doctrine to foresee accurately the nature of the coming air war. Such arguments miss the critical point that American doctrine, when combat experience was directly integrated into its conceptions, helped provide a strategy for the employment of air power that played a major role in winning the Second World War. Basic American doctrine, evolved in the 1930s, drove the requirements for technology, set the goal for a great bomber fleet, and when applied with flexibility and imagination (as in the Mediterranean in 1943 and from England in 1944), provided a framework within which the Army Air Forces could adapt to the realities of air war against the Third Reich. Some air commanders, however, refused to adapt prewar doctrine to the realities. The results were catastrophic for the crews flying combat missions. Unfortunately, the commanders of Eighth Air Force in 1942 and 1943 held to many of their doctrinal assumptions of the prewar period. Thus, while there is little need to search for an explanation for the sixty bombers lost in the Schweinfurt/Regensburg raid in August 1943, the second great disaster over Schweinfurt in October 1943 (where another sixty bombers were lost) raises serious doubts about the willingness of Eighth's leadership to adapt doctrine to reality. One of the major lessons of the Second World War was that doctrine was not written in stone. A military force unable or unwilling to adapt would suffer the consequences. Of course, even worse than a military force unwilling to adapt doctrine was a force with no doctrine at all. For the Italian military in 1940, the consequences were complete, crushing, and absolute—so absolute that Italy's military forces failed to recover for the remainder of the war.[6]

Given the importance of doctrine to the conceptual preparations of military forces for war, this chapter will examine the doctrinal statements of two air forces not only for their realism but their adaptability. The first statement comes from the 1930s, and enunciates the clearest vision, for the time, of how air power would be used in the Second World War. It is an especially clear statement on the difficulties that air forces would face in the conduct of air operations during future conflicts. The second doctrinal statement came in 1979 from the air force of one of the world's strongest military powers and supposedly guided that military organization until 1984. It had a wealth of experience on which to draw—nearly all of which was not available in the 1930s. What it discussed, and how it cast its doctrine serves to underline an enormously different *Weltanschauung*. It suggests that the arguments of those who have written about the loss of professional military virtues in the armed forces of the United States have hit all too close to the mark, at least for the period when the manual appeared.[7]

The basic doctrinal statement of the *Luftwaffe* appeared in the year of its birth, 1935. Among the authors was the new service's brilliant chief of staff, Walter Wever, who had been one of the developers of German defense doctrine in the First World War. He was so highly regarded in the early 1930s that Hitler's defense minister, *Generaloberst* Werner von Blomberg (later field marshal), had Göring select one of two senior army staff officers for the position of Luftwaffe chief of staff: the two were Wever and the future Field Marshal Eric von Manstein, arguably the best operational general of World War II. Göring selected Wever.

Wever's doctrinal manual was entitled simply, "The Conduct of the Air War (*Die Luftkriegführung*)".[8] The title underlined in the clearest fashion that this manual was about the conduct of air war and repetitively throughout the manual one runs across words like war, combat, battle, and the offensive. The second theme was the importance of leadership: leadership from the front, leadership by "personal example, [by] the appearance of leaders at the places of greatest danger." While higher ranking leaders would not be able because of the nature of modern war to lead in battle, the manual suggested that it was incumbent on them to utilize every opportunity for personal leadership.[9]

The first section of the manual discusses the Luftwaffe's place within national military strategy. As with other air power theories of the interwar period, the introduction suggested that the Luftwaffe had the possibility to "carry the war from the beginning against the enemy's home land [and to] attack his military power and the morale of the enemy population at the root."[10] Nevertheless, in clear divergence from the air power theories of the time, the authors go on to argue that even though the Luftwaffe might operate in a different medium than the army and navy, it must never lose its sense that air force operations were only a part of a larger whole and that it was only one of *three* national military services.[11]

The clearest statement of the strategic purposes for the Luftwaffe comes in the manual's opening paragraphs:

> 9. The task of the military (Wehrmacht) in war is to break the enemy's will. The will of a nation finds its strongest embodiment in the military. Therefore, the destruction of the enemy military is the foremost goal of war.

10. The task of the Luftwaffe is this: Through its conduct of the air war within the framework of overall strategy to serve this goal. Through battle against the enemy air force [the Luftwaffe] weakens the enemy military and at the same time strengthens one's own military, the population, and [one's] *Lebensraum*. Through intervention in operations and combat on land and at sea the Luftwaffe supports the army and navy. Through attacks against the sources of support for the enemy's military forces and the support structure from industry to front the Luftwaffe seeks to weaken the enemy's military.

11. How to determine what strategy will have the strongest effect, what tasks should be placed first in the order of air force priority, can only be determined within the framework of the overall military situation . . . The enemy, the weather, the time of year, the structure of the [enemy] land, the popular character and one's own military power must be considered in [strategic] decision making, in order to determine the possibilities and limits of the Luftwaffe.[12]

This lengthy quote underlines certain critical points in evolving a theory on the employment of air power. To begin with, the statement makes clear that "strategic" bombing played a considerable role in Luftwaffe thinking.[13] But "strategic" bombing is placed squarely within the context of overall national strategy and beside other possible employments of air power. Moreover, this manual states clearly what the role of air power is in warfare and nowhere does it confuse the classical definitions of strategy, operations, and tactics with what Anglo-American air power theorists have misnamed "strategic" bombing. The difficulties into which the division of air power missions into "strategic" and "tactical" bombing have led us were underlined by a debate that this author had with an American air force officer in 1980. The argument was over whether one could term the German breakthrough on the Meuse in May 1940 as a strategic or tactical victory. This led him to examine the question of air power employment within that campaign more closely and to frame the following question: What set of missions would have enabled the Luftwaffe to further overall German strategy best in May 1940? Supporting the army's breakthrough efforts along the Meuse or bombing France's industrial base and cities? In the classical definition of strategy and tactics the answer is clear. In terms of Anglo-American air power theories the question and answer are not so forthcoming.

Thus, Luftwaffe doctrine had no difficulty in placing "strategic" bombing within the context of an overall national strategy. The results were a more realistic appraisal of the prospects for "strategic" bombing. The manual argues that the battle against the enemy's sources of economic power can have decisive importance. Such an offensive would attack the enemy at the root of his will to resist. "But it [such a campaign] also takes," the manual warns, "a long time to take effect and therefore carries the danger that it would come too late to influence the struggle on land and at sea."[14]

Unlike the air power theories in Britain, the United States, and Italy in the 1930s, the Luftwaffe manual argued that the threat to one's own land and military services from enemy aircraft demanded the employment of air power to seek out and attack enemy air forces from the beginning of a war. Such an air offensive necessitated not only an attack on enemy air units and bases but on the production facilities and industrial plant that supported the enemy air force. Surprisingly, considering that non-German air theorists had almost entirely bought the "short war" scenario, Luftwaffe doctrine recognized the difficulty not only in gaining but maintaining air superiority. "One must also calculate on only a temporary air superiority, because the enemy can within a short period of time replace his losses, or regain operational freedom by technical changes."[15] While one can doubt the claim that production or technological changes could radically change the air balance of power in a conflict,[16] the appraisal was more realistic on the elasticity and long-term staying power of modern military institutions than the theories of Trenchard, Douhet, or the Air Corps Tactical School in the U.S.

"The Conduct of the Air War" then turns to the subjects of leadership, employment, tactical and operational strategies, interservice cooperation, conceptualization of an air offensive, and defensive operations. For our purposes, we need only note the generally realistic, broad-based approach to the employment of air power. The arguments and discussions hold up surprisingly well despite being written over fifty years ago. Worth re-emphasizing is the overall impression that the manual conveys: that it was written for a force *preparing for war*. Nevertheless, there is one glaring weakness that played a crucial role in the Luftwaffe's defeat in World War II. There are no sections in the 1935 manual that deal with ground organization, intelligence, communications, supply, or production. Those sections, the final page tells us, were still being written. They were never completed.

Where then did the Luftwaffe fail if its doctrinal manual gave such clear direction on the employment of air power? To begin with, the manual represented the strategic foresight of one airman, Wever, who was killed in 1936 leaving no successors with his sense of the larger issues. Even had Wever lived, it is uncertain how long he would have contributed given the nature of Hermann Göring's leadership. Certainly Wever's belief that there was a relationship between means and ends would would not have pleased the Reichsmarschall and the Führer.

While the Luftwaffe in the early 1940s proved *the* most capable air force at directly influencing (along with the army) the European balance of power, those in leadership positions believed megalomania to be the equivalent of national strategy. On the other hand, the doctrinal framework, within which the Luftwaffe fought, helped to make it an extraordinarily tenacious opponent.

Rumor has it that the 1979 version of United States Air Force Manual 1-1, "Functions and Basic Doctrine of the United States Air Force" was written to "tell the Air Force story." No matter how commendable such a goal might be, that is not the purpose of a doctrinal manual preparing a military force and its commanders for war. In 1984 a new version appeared and at this writing (1990) another revision is in the works. Nevertheless the 1979 version is worth examining for it reflects trends that dominated the air force at the end of the Carter Administration and were not entirely eradicated during the Reagan years.

The drawings of aircraft along the margins of the text, the pictures of notables and commentators on air force doctrine such as President Jimmy Carter, Chief Master Sergeant Robert D. Gaylor, and Major-General Jeanne M. Holm, and the graphics of "managing people" ("PROCUREMENT, get good people, EDUCATION AND TRAINING, prepare them, UTILIZATION, use them properly, SUSTAINMENT, support them, SEPARATIONS AND RETIREMENTS, separate them") created more of a public relations tone than one required for a manual discussing matters which in the final analysis involve life and death.[17]

Internal contradictions also detract from the manual's effectiveness.

The first perhaps reflects beliefs of earlier air power theorists that aircraft had invalidated the principles of war. Chapter 2 opens with the statement that the fundamental purpose of the air force is to "conduct prompt and sustained combat operations in the air to defeat enemy airpower."[18] Yet in chapter 3 we discover that the characteristics of air power "enable the direct application of power against all elements of an enemy's military resources to a degree not possible by other forces."[19] What little sense this latter claim makes in the face of air power's inability to seek out guerrilla forces moving through Vietnam's jungles is only reinforced by the section on unconventional warfare that reads as if that conflict had never occurred.[20]

Throughout the manual there is an emphasis on the role of the USAF in deterrence as opposed to combat. In fact, chapter 1 places the role of military force in sustaining deterrence above the conduct of war.[21] On the national level this undoubtedly reflected the basic assumption on which many of America's defense policies rested in 1980. Nevertheless, when a nation's military services become more concerned with deterrence than with the business of *fighting*, their ability to deter is significantly affected.

Failing to come to grips with the air force as a combat force, the manual instead spends much of chapter 3 discussing the management of air force personnel ("Managing People is a Dynamic and Complex Task"). Clearly the emphasis is on the "civilianization" of the military and the turning of the Air Force into just another profession. The manual tells us:

> Military leaders must show concern for their people—for their quality of life. Each superior at every level should ask, 'What have I done today for the people who work for me?' Personnel policies must insure equity, opportunity, and reward. . . . All of our people must be fully aware of matters involving their day to day participation in military operations and their living environment.[22]

As for the subject of leadership—the art of inspiring individuals and units to fight together under the terrible conditions on the modern battlefield—the manual has virtually nothing to say.

The 1979 AFM 1-1 shows great concern for tightly centralized command and control systems. It suggests that the AWACS (Advance Warning and Control System) "allows commanders to *comprehend the total air-surface battle*" (emphasis added).[23] Perhaps the statement has some validity in a brushfire war, but not to a central European or Middle Eastern battlefield. In great detail chapter 4 lays out the command and control system of the USAF (Unified Command Structure, Joint Force Organization, Air Force Component Command, Command, Coordination Control, and finally The Control System). The wiring diagrams reflect the worst aspects of the German love for military minutiae,[24] although in reality the Germans paid little attention to such diagrams. Again leadership plays no role in determining what to do when communications freeze up or break down, when headquarters are bombed out, or when commanders are on the run.

Chapter 5 does address several basic issues. While centralized control receives pride of place, the value of decentralized execution also receives attention.[25] But how the authors hoped to integrate such divergent concepts is not discussed. Unfortunately, American operations in Vietnam or even in the Iranian rescue mission hint as to where the emphasis lay in the 1970s. The rest of chapter 5 then turns to a discussion of the "principles of war" and is a dangerous over-simplification of the complexities, ambiguities, and chaos of combat. The interest in Clausewitz in the American war colleges since the Vietnam war (particularly at the Naval War College) underlines the lessons of the past twenty years—that simple schoolbook solutions are never sufficient. The manual concludes with a discussion of the evolution of basic doctrine in the USAF and in its predecessor organization. Again the authors are unwilling to grapple with fundamental issues. There is no mention of the failures of 1930s doctrine applied to 1943 conditions and little discussion of the difficulties involved in the great "strategic" bombing campaign. Instead according to the manual "the destructive power of strategic forces was generally immediate and overwhelming."[26] What the crews who went back to Schweinfurt a second time in October 1943 might have thought of such a simplification we have little chance of finding out, since most died on that mission or on missions shortly thereafter.[27]

Several final comments are in order. First of all, is the glaring ommission of the Vietnam War, which is never discussed directly in the manual. Only an oblique reference in chapter 4 suggests that an edition of AFM 1-1, written during the 1964-1971 time frame, was put together "during an unpopular conflict that was not, and is not yet, clearly understood."[28] The manual also consistently uses buzzwords like "warfighting," instead of the words which have proved more than sufficient for military historians and theorists in past ages, such as war, fight, combat. The question of doctrine, of course, represents only the first step toward the creation of effective fighting forces. A decentralized adaptable doctrine demands constant training that emphasizes subjective rather than objective criteria in judging peacetime military performance. In peacetime training such a system must force military units to face new and unforeseen circumstances demanding adaptability and flexibility. In other words, the criteria used to judge unit effectiveness cannot be dull or repetitive training, but rather exercises that challenge the inventiveness of commanders and subordinates—failure cannot be judged on the basis of narrow confining standards that have little relationship to possible combat situations. Subjective standards are more difficult to judge; and in organizations that place maximum emphasis on objective measurements there will be those who claim that subjective criteria cannot be an effective management tool.

Unfortunately, as the events of 1990 have shown, peace in one part of the world does not remove threats in others. The circumstances and conditions of war rarely resemble what military organizations have prepared to meet. The authors of AFM 1-1 suggested that the framework within which they cast their effort is one that was "descriptive in nature and should be viewed from a philosophical, not a legal, context."[29] That is indeed the proper approach. The problem is that 1979 AFM 1-1 was not a doctrinal statement. It did not possess the broad approach of the Luftwaffe's "Conduct of the Air War" nor did it even contain the tight framework within which the Army Air Forces prepared in the early 1940s. As a result, with hardly any frame of reference for the employment of air power under wartime conditions, one suspects that USAF commanders of 1980 would have had a difficult time in adjusting to the unexpected; and warfare, even in the age of advanced electronics and computers, has shown itself to present continuously new and unforeseen conditions.

NOTES

1.   For one of the most interesting recent studies on the importance of doctrine to the creation of effective military forces, see Timothy F. Lupfer, *The Dynamics of Doctrine: The Changes in German Doctrine During the First World War* (Leavenworth, 1981).

2.   For a discussion of what hard realistic training can contribute to the combat readiness of military forces, see my article in *Armed Forces and Society*, "German Response to Victory in Poland: A Case Study in Professionalism" (Winter 1981).

3.   The German attitude might best be summed up by the aphorism "better no officer than a bad officer."

4.   For a discussion of the development of American air doctrine in the late 1930s and early 1940s consult Thomas H. Greer, *The Development of Air Doctrine in the Army Air Arm, 1917-1941* (Montgomery, 1955); Thomas A. Fabyanic, "A Critique of United States Air War Planning" (Ph.D. Dissertation, St. Louis University, 1973); Haywood S. Hansell, Jr., *The Air Plan That Defeated Hitler* (Atlanta, 1972); and Robert F. Futrell, *Ideas, Concepts, and Doctrine*, vol. I (Montgomery, 1971).

5.   For a recent discussion of the American attack on Schweinfurt, see Friedhelm Golücke, *Schweinfurt und der strategische Luftkrieg 1943* (Padeborn, 1980).

6.   For a brilliant discussion of the inability of Italy's military forces and commanders to articulate either doctrine or strategy, see MacGregor Knox, *Mussolini Unleashed, 1939-1941* (New York, 1982).

7.   See among many articles and books the outstanding article by Frank R. Wood "Air Force Junior Officers: Changing Prestige and Airbornization," *Armed Forces and Society* (Spring 1980).

8.   "Die Luftkreigführung," (Berlin, 1935); copy made available to the author by Oberstleutnant Klaus Maier of the Militärgeschichtliches Forschungsamt, Freiburg, Federal Republic of Germany.

9.   Ibid., paragraphs 6 and 7.

10.   Ibid., paragraph 2.

11.   Ibid., paragraph 8.

12.   Ibid., paragraphs 9, 10, and 11.

13.   For a more detailed discussion of the role of "strategic" bombing in the Luftwaffe's thinking, see chapter 2.

14.   "Die Luftkriegführung," paragraph 22.

15.   Ibid., paragraphs 16, 17, and 18, and especially section IV, paragraphs

103, 104, 105, and 106.

16.   See chapter 3 of Williamson Murray, *Luftwaffe* (Baltimore, 1985).

17.   Air Force Manual 1-1, chapter 3, p. 6.

18.   Ibid., chapter 2, p.1.

19.   Ibid., chapter 3, p. 2.

20.   Ibid., chapter 2, p. 19.

21.   Ibid., chapter 1, pp. 7-11.

22.   Ibid., chapter 3, p. 8.

23.   Ibid., chapter 2, p. 22.

24.   See especially the diagram on p. 8 of chapter 4.

25.   Ibid., chapter 5, pp. 1-8.

26.   Ibid., chapter 6, p. 2.

27.   For the implications of attrition, see Murray, *Luftwaffe*, chapter 5.

28.   AFM 1-1, chapter 6, p. 5.

29.   AFM 1-1, p. vii.

CHAPTER
SIX

THE LUFTWAFFE AND
CLOSE AIR SUPPORT 1939-41

Since aircraft became a weapon of war in the early twentieth century, close air support has played a major role in the employment of air power. This chapter sets out how the Germans began the development of a close air support doctrine and capability in the years immediately before World War II and then how they refined those concepts as the conduct of operations and battlefield conditions suggested employment possibilities.

Air-ground cooperation on the immediate battlefield has never been an easy matter to orchestrate and neither the Luftwaffe nor the German army found it easy to work out such operational concepts. What is of interest, however, is the relative openness with which German ground and air officers approached the problem and the relative lack of rancorous debate that accompanied the evolution of common doctrine and concepts. When the war began in September 1939, the Luftwaffe still lacked satisfactory methods for aiding the army with direct, close air support missions. Nor were most air force commanders convinced that this role represented the best employment of air power, though they were willing to approach the problem with an open mind.

The traditional picture of German victories during the 1939-41 period depicts a combination of tanks, infantry, and Stukas working in close and explosive cooperation to overwhelm the hordes of World War I-type infantry which the other European powers placed on the battlefield to oppose the Reich's advance. Like much of military history, this picture has both exaggeration and truth. While it is clear that at certain critical moments, especially along the Meuse between 13 and 14 May 1940, close air support greatly contributed to German success, evidence suggests that at least in the early war years close air support for the army's advance played a relatively small role in the Luftwaffe's operations. As with most other air forces in the 1930s, the Luftwaffe was only beginning to evolve a system of army-air force cooperation that could be called sophisticated.[1]

As rearmament proceeded in the late 1930s, the Germans had considerable experience on which to draw from World War I. In the trench stalemate of 1914-17, aircraft contributed photo reconnaissance, interdiction, and even close support to frontline troops. That stalemate with its clearly defined opposing trench systems provided a static environment within which reconnaissance, fighter, and ground support aircraft could render important help. By the time of the Flanders battles in 1917, the Germans had evolved a system where air liaison officers served with frontline divisions while radio technology provided communication between air observers and frontline artillery batteries.[2]

As the system of air-ground cooperation was evolving, changes in German offensive doctrine at the end of 1917 posed major problems for the support of ground forces from the air. In effect, the German general staff managed to design and implement an infantry doctrine that returned maneuver to the battlefield.[3] That revolution in operational concepts and capabilities, however, meant that once German armies had broken through the enemy front lines and had reached the exploitation phase, communication and coordination between air units and advancing ground forces would become more and more difficult. The Germans recognized this, and the great German ace and operational commander, Manfred von Richthofen, devoted a section of his lessons of the air war to support for the army in "breakthrough battles and maneuver warfare (*Bei Durchbruchsschlachten und Bewegungskrieg*)."[4] Complicating the transition was the fact that a major reorganization of signal troops worked to the disadvantage of air units supporting the spring offensives.[5] Despite Germany's progress toward a more effective system of ground cooperation in 1918, that system was still primitive. Several factors were clear: along with effective communications, air-ground cooperation depended on general air superiority. Here, overwhelming Allied air strength made it increasingly difficult for German air units to intervene in the ground battle as the year progressed.[6]

In 1919 the Treaty of Versailles removed aircraft from the German inventory of weapons for the next fourteen years, but there was some experimentation in Russia between the military of the Weimar Republic and the Soviet Union. Moreover, Hans von Seeckt, creator of the postwar German army, saw to it that a small but significant number of officers with flying experience remained in the tiny postwar officer corps. These factors, however, created a situation in which most officers had virtually no experience with aircraft. With Hitler's coming to power in January 1933, the Germans underwent a rapid military expansion that took them from the depths of disarmament to the buildup of forces that destroyed the European equilibrium in 1940.[7] During that process of rearmament, the creation of an effective and powerful air force was crucial to German success in the early years of the war.

The traditional picture that the Luftwaffe was "in effect the hand maiden of the army"[8] largely distorts the intentions and strategic vision of those who created the Luftwaffe. At the beginning of the war most of the Luftwaffe's high command and officer corps believed in the importance of "strategic" bombing. Moreover, they believed that the Luftwaffe was on the road to creating the force structure required to make it an effective "strategic" bombing force, at least within Central Europe.[9] Other air forces and most European statesmen (as well as Germany's leaders) in the late 1930s agreed. Nevertheless, if the Luftwaffe's leaders were pushing toward the creation of a "strategic" bombing capability, they also placed the Luftwaffe's strategic conceptions within a broad framework of national strategy and interservice cooperation. Such attitudes were in distinct contrast to those of most air power theorists (military as well as civilian) in Great Britain and the United States.

The Luftwaffe's first chief of staff, General Walther Wever, played a critical role in the development of German prewar air doctrine. Wever possessed one of the best operational minds among his generation of officers (Blomberg offered Göring the choice between Wever or the future field marshal Erich von Manstein for the position of the Luftwaffe's first chief of staff).[10] Unlike many of his German army colleagues, Wever

possessed a generally realistic understanding of the relationship between operations and strategy.[11] This understanding gave him a keener
appreciation of the political and strategic context within which the
Luftwaffe might fight than was the case with other theorists such as
Douhet, Trenchard, or most of those in the American Air Corps Tactical
School. Thus, Wever was anything but an unabashed champion of
"strategic" bombing. As he made clear in 1935 at the *Luftkriegsakademie*
(Air War College), the Luftwaffe's status as a separate service did *not*
mean that its employment would be independent of the army or the
navy. Rather, depending on circumstances, its contribution could involve missions as varied as attacks on the enemy air force, army, fleet,
and industrial base. The goals and purposes of national strategy would
play the critical role in determining air power employment.[12]

The clearest statement of Wever's conception of air war came with
the publication of the Luftwaffe's doctrinal manual in 1936 ("*Die
Luftkriegführung, the Conduct of the Air War*").[13] In it Wever and his co-
authors showed a ready grasp of the political and strategic complexities
of twentieth century warfare. They clearly understood that air war
would be inseparable from the conduct of campaigns in other dimensions. They recognized first that air superiority would be a critical but
elusive goal. "Strategic" bombing, while a major factor, represented an
unknown quantity and might well take too long to be decisive, at least
from the German perspective.

On air-ground cooperation, "Conduct of the Air War" was explicit
in its argument that the Luftwaffe could and should aid the Reich's
ground forces. It warned that there would be difficulty in close cooperation in finding targets "against which [bomber units] could bring their
full attack potential to bear."[14] Nevertheless, the manual suggested that
the Luftwaffe should be committed in support of the army in critical
moments of a land battle. As to when and where, the manual argued
that the basic requirement would be that the mission must produce
decisive results for the success of ground operations: "The closer the
contesting armies are locked in combat and the closer the decision in
battle comes, the greater will be the effectiveness of bomber attacks in
the battle area." However, it pointed out that close air attacks against
well camouflaged enemy forces in good tactical positions were "unlikely
to produce results commensurate with the effort." Moreover, air attacks
on enemy forces within range of friendly artillery fire should only occur
where the weight and capabilities of artillery were insufficient for the

mission.[15] The manual creates the impression that while close air support was an important mission, it was of subsidiary importance to missions such as interdiction, the achieving of air superiority, and, in certain cases, "strategic" bombardment aimed at enemy industrial or economic resources. Unlike the RAF, which generally rejected the close air support mission except in the most desperate of circumstances, the Luftwaffe was willing to look on the close air support mission as one in which its forces should be employed on a sustained basis. From a Luftwaffe perspective, it did not necessarily represent the best employment of air power in general terms, but it was a mission in which air power could and should render significant help to the ground forces when the overall battlefield situation demanded it.

The development of the Luftwaffe between 1933 and 1939 followed priorities Wever established. The prewar emphasis remained on creating an interdiction and "strategic" bombing force. To that end the Germans created the largest bomber force in the world. Admittedly, their aircraft were only twin engine bombers, but these were considered sufficient for "strategic" bombing attacks within Central Europe. Meanwhile, Luftwaffe engineers were hard at work on a four engine aircraft, the He 177, with the range and payload to attack more distant targets.[16]

The German post-1933 rearmament effort faced considerable obstacles in the aircraft sector. To begin with, the Reich's aircraft industry numbered approximately 4,000 workers scattered among a number of underutilized and undercapitalized firms.[17] Nevertheless, priorities were clear from the start: first, the establishment of a "strategic" bomber force; second, an air superiority fighter force; and third, an antiaircraft artillery capability that could defend German industry from enemy bombing attacks.[18]

While the Luftwaffe did believe that air support for the army could be a major role in a future war, little preparation was done in early rearmament years. The individual problems confronting the services during rapid expansion were daunting enough. The army's emphasis through 1938 remained on creating a well-trained World War I infantry force. Hitler did not interfere in the buildup, and while armor advocates such as Guderian and Lutz were creating the kernel of the future panzer force, their efforts remained on the sidelines in the overall rearmament picture.[19] There were some contacts between the Luftwaffe and the new panzer divisions and one former Luftwaffe officer participated in a joint army air force command post exercise in 1936 that also included Guderian as a participant. But he remembers the panzer general as having been generally unrealistic and unknowledgeable as to the capabilities and limitations of aircraft support.[20]

The Luftwaffe's experience in the Spanish Civil War played a critical role in pushing it toward a more accurate assessment of its equipment as well as providing it with a modicum of air combat experience. Air to air tactical experience was particularly important. Unlike the Italians, the Germans limited their commitment in Spain to a small and manageable size.[21] The initial contribution to Franco's cause came when Ju 52s ferried Nationalist troops from Morocco to Spain. The German combat aircraft first deployed to Spain, the bomber version of the Ju 52 and the He 51 fighter (both aircraft representing the first generation of aircraft production), quickly proved inferior to Russian aircraft on the Republican side.[22] In fact, this clear inferiority may have been *the* major contribution of the Spanish Civil War to German rearmament. It forced the Germans to send out their new prototype fighters and bombers and shifted German production quickly to second generation aircraft.

At its height in the autumn of 1938 the Condor Legion (cover name for German air aid to Franco) comprised only forty He 111s, five Do 17s, three Ju 87s, forty-five Bf 109s, four He 45s, and eight He 59s.[23] But from that force the Germans learned important doctrinal and technological lessons. By late 1938 the fighter ace Werner Mölders had developed the finger four formation that all air forces would eventually adopt but which gave the Luftwaffe a considerable edge in the first encounters of World War II.[24] The Condor Legion also made the first steps toward development of a close air support doctrine. The leader in this area was Wolfram von Richthofen, cousin of the great World War I ace. Richthofen arrived in Spain out of favor with some in the Luftwaffe's high command but with a similar frame of reference to those in the air ministry that he had just left: close air support was a low priority. However, once Richtofen became the Condor Legion's chief of staff, he recognized that theoretical musings on "strategic" bombing and the political and military realities of the Spanish Civil War had little in common. The stalemate on the ground, the lack of suitable targets for strategic bombing attacks, the weakness of Nationalist artillery, and the combat deficiencies of the first German aircraft led him to push for employment of available air power in direct support of Franco's offensive against Bilbao.[25]

With little official encouragement from Berlin, Richthofen developed a primitive, but for its time, effective close air support doctrine and capability.[26] Before the Bilbao offensive launched by Franco in 1937 against the northern Spanish port, few of the tactical or support elements required for close air support existed in German doctrine or practice. But within a year, the Condor Legion had evolved a system that insured close planning between ground and air units, established communication links and recognition devices, and detailed Luftwaffe officers to serve directly with frontline units. Significantly, there was not much enthusiasm for the experiments among the Luftwaffe staff in Berlin.[27] In retrospect, Richthofen had really only managed to reintroduce German close air doctrine as it existed at the end of World War I. Recognition devices, liaison officers, telephone and radio communications had all been used during the 1918 spring offensives. One of Richthofen's close associates in Spain, Major General Hans W. Asmus, suggests that Richthofen had not innovated but rather had drawn from his own wartime experience and those of others to establish the procedures for air-ground cooperation on the battlefield. In some respects the Condor Legion's system was even more primitive than that of World War I; in some cases German pilots identified Spanish infantry by the flags they carried.[28] In other cases Nationalist infantry wore large pieces of white cloth on their backs, thus making advancing troops easy to spot from the air and at the same time discouraging thoughts of retreat.[29]

What Richthofen could not solve was the problem of coordinating close air support strikes with the rapid exploitation drives of motorized and mechanized formations. That experience was not attainable in Spain, because military operations closely resembled those of World War I—with infantry breakthrough operations against static defense lines providing the basis of combat. The tactical and operational concepts of Nationalist military leaders as well as the capability of their armies simply did not allow for rapid mobile operations. Moreover, the Spanish forces possessed primitive communication links, especially in terms of radios, so that telephone links represented the most advanced communications available.

Interestingly, the Ju 87, the famed Stuka, flew few close air support missions in Spain. Some Stukas were sent to Spain (only three were there in autumn 1938), but those that did go were sent for combat evaluation against precision targets such as bridges, railyards, and other choke points. The air staff in Berlin regarded close air support missions as too dangerous for these planes.[30] But the Stuka proved both its sur-

vivability in attacks on interdiction targets as well as its superior bombing capability compared to conventional horizontal bombers such as the
He 111 or the Do 17. With constraints on ammunition production due
to lack of industrial capacity in Germany in the late 1930s, it was the
Stuka's accuracy that made it so attractive and that led Ernst Udet,
director of the Luftwaffe's design bureau, to push for a dive bombing
capability for future German bombers no matter how big—a disastrous
mistake.[31]

By 1938 Richthofen's experiments in Spain had created a place for
close air support in the Luftwaffe's preparations—but one that was still
relatively low in terms of the Luftwaffe's other missions. The
Luftwaffe's aircraft inventory at the outbreak of war against Poland
underlines its priorities: 1,180 bombers, 771 fighters, and 366 dive bombers. Only dive bombers had the trained crews and capability to perform
close air support and the Stukas also had to support air interdiction
missions and attacks on the enemy air base structure.[32] Moreover, the
one wing of close air support aircraft possessed obsolete fighters and
was assigned the mission because of the unsuitability of its aircraft for
any other role. Significantly the Germans made no effort to design a
primary mission aircraft for supporting the army in the battle zone.[33]

As the Spanish Civil War burnt itself out in 1938 and 1939, the
Germans began final preparations for what turned into World War II.
While interservice cooperation generally was good, particularly at infor-
mal, lower command levels, considerable conceptual differences influenced the extent to which the Luftwaffe would directly support the
army. At a May 1938 wargame one participating Luftwaffe officer underlined those important differences in a memorandum.[34] He attempted to
clarify for the participating army officers that the Luftwaffe's primary
goal in war would be the destruction of opposing air forces. In the case
of "Fall Grün" (war with only Czechoslovakia) that period of time would
probably last four days; in the case of "Fall Rot" (war with France) at
the minimum, four weeks. Only then could the Luftwaffe support the
army with its bomber squadrons (in interdiction as well as close air
support missions). The army officers' reply, irrelevant on the issue of
air superiority, was that Spain had shown air support of ground operations to be more important than any effects gained by "strategic" bombing. The Luftwaffe's representative reported that "army officers again
and again uttered the desire to employ the air force on the battlefield
and for this purpose to support each army with a bomber Geschwader"

(squadron).

In May 1939 the major army general staff exercise again suggested considerable differences between the views of Luftwaffe and the ground forces. A Luftwaffe staff paper stressed that in no case would bomber units be placed directly under army control.[35] Rather the army must state its priorities, requirements, and requested time for support to the appropriate Luftwaffe command level which would then determine what it would and could support. There would be *no* employment of bomber units in the immediate battle zone. Major General Hans Jeschonnek, Luftwaffe chief of staff in 1939, emphasized the air force's position. The Luftwaffe's emphasis in any future war must first of all be on the destruction of enemy air forces and the achievement of air superiority.[36] On close air support for the army, Jeschonnek suggested that this was a difficult task about which much was unclear. He then compared it to a cavalry charge: "It could bring great success when it achieved surprise but only then. When it did not possess surprise and met an enemy who was prepared, then such an attack had little success and that at a disproportionally high cost." Jeschonnek expressed considerable doubts as to the effectiveness of close air support materially as well as morally, especially against a first class enemy.

Before turning to the evolution of close air support doctrine in the first years of World War II, it is important to examine the organizational and communication ties between army and air force in the late 1930s that coordinated employment of airpower in support of ground operations. At the highest level, Göring assigned a *General der Luftwaffe beim Oberbefehlshaber des Heeres* (Luftwaffe [liaison] general to the commander in chief of the army).[37] Unfortunately for the Germans and typical of Göring's brand of leadership, this liaison officer possessed no authority to discuss common army-air force problems. Rather he served two distinct functions: he was Göring's messenger to the army high command; and he commanded the close and long range reconnaissance aircraft assigned directly to army support (approximately 450 aircraft at the beginning of the war).[38] Thus, doctrinal differences and tactical problems between the army and air force were not funneled through one liaison office but when faced, if at all, were addressed on an ad hoc basis at different levels of command.

The reconnaissance squadrons, directly assigned to the army, while not strictly falling under the rubric of close air support, do pose interesting questions about the system of air-ground cooperation and suggest considerable systemic weaknesses on the outbreak of war. Under the *General der Luftwaffe beim Oberbefehlshaber des Heeres* were *Kommandeure der Luftwaffe* (*Kolufts*) (commanders [or officers] of the Luftwaffe) assigned to army groups and subordinates who in turn commanded Kolufts at army level. These Kolufts were responsible for reconnaissance squadrons assigned at corps level (infantry as well as mechanized and motorized). The chain of command for close air reconnaissance ran down to the army corps level (through the French campaign close air reconnaissance squadrons were only rarely assigned to panzer divisions).[39] There was some organizational confusion because the *Kolufts* at the various levels could order reconnaissance squadrons to support other army groups or armies without the knowledge of army authorities.[40] Nevertheless, the system was generally effective, although costly in terms of aircraft and crews. Radio communications ran directly from He 126s to artillery batteries on the ground (in morse code transmissions through the end of 1939 and by voice at the opening of the French campaign).[41]

A second system of coordination between army and Luftwaffe was through *Fliegerverbindungsoffiziere*, air liaison officers, or *Flivos*, for short. These officers were assigned by the numbered air forces (*Luftflotten*) to army groups, by *Fliegerkorps* to armies and by air divisions and *Geschwaders* (squadrons) to army corps. The Flivos played a critical role in coordinating and informing operational air and ground units of what the army required and what the Luftwaffe could or would supply. Surprisingly there was no direct relationship between the *Kolufts* and the *Flivos* except on the personal level—a major weakness, with coordination between the Luftwaffe offices cooperating with the army being done informally. Moreover, and here lay the greatest weakness in the system, there was no means to communicate directly between the close reconnaissance squadrons and Luftwaffe fighter, bomber, or dive bomber units.[42] There were no common radio frequencies between the army and Luftwaffe units. But cooperation did in fact work somewhat better than the organizational outline might suggest—largely because of a willingness of those at different levels of command and service to pull together despite organizational differences or limitations.[43] But it does seem surprising that the *Kolufts* were virtually excluded from the coordination process. The only satisfactory explanation appears to be Göring's desire to keep the entire decisionmaking process within the Luftwaffe under his personal control.

In assessing the prewar system several points are significant. The system possessed serious weaknesses in organization, coordination, and tactics. The close air reconnaissance squadrons were excluded from the close air support loop despite the fact that they could have rendered direct, effective help to the coordination and direction of close air support strikes. Nevertheless, despite such systemic weaknesses, the Luftwaffe was one of the few air forces in the late 1930s that had thought about the problems of close air support and had recognized that an air force could render important help to ground troops in critical situations. The system was best at providing help for the army when it assaulted well-defined enemy defensive lines—before the armored forces began the process of exploitation. It was *not* effective at supplying close air support once panzer units were in the open and moving with the rapidity that caused such surprise and consternation among the other European armies.

For the attack on Poland the Luftwaffe set for itself three basic missions: destruction of the Polish air forces, their ground service organization, and the Polish air armament industries; support of army operations on the ground in order to insure a quick breakthrough on the ground and a speedy advance by the ground forces; and attacks against Polish military installations and armament industries in Warsaw.[44]

How the Luftwaffe executed these three missions in the Polish campaign is important. "The Luftwaffe did not go in for the 'tidy priorities' beloved of the AAC [Army Air Corps] and the RAF, both of whom were following a policy that was political rather than operational and could only be defended if doctrinal roadblocks were placed in the way of those who tried to insist" that ground operations could be decisive.[45] Rather the Luftwaffe set a general mission framework for itself and then executed its air campaign in accordance with the realities of combat, the conduct of the war on the ground, and its logistics and operational capabilities. In other words, it adapted to the real conditions of combat as fast as it could. In its plans the Luftwaffe scheduled a major raid on Warsaw for the early morning hours of 1 September for the opening move of the air war. Weather conditions, however, prevented execution of this operation (clearly an effort at "strategic" bombing).[46] Consequently, Luftwaffe operations in Poland emphasized its two other basic missions. As prewar doctrine had suggested, the Luftwaffe high command regarded the achievement of air superiority as its major operational goal. In the first days of the campaign the Luftwaffe's major emphasis was on strikes against the Polish air force. At the same time ground support Hs 123s with some Stuka support helped the army break through Polish defenses and achieve the operational freedom that panzer and motorized units required in order to execute deep penetration and exploitation drives. By the fourth day of the offensive, German mechanized forces were loose and rampaging through Polish rear areas. The Luftwaffe did render some direct support to these units, but the bulk of its sorties were interdiction strikes against the transportation system and direct strikes against a crumbling Polish army. Particularly along the Bzura River, the Luftwaffe struck so effectively against Polish army units attempting to regroup and counterattack German breakthroughs that some Polish troops simply threw away their weapons.[47]

The devastating nature of the Wehrmacht's success drew an interesting response from the German military. The army high command, despite having destroyed enemy armed forces of thirty-plus divisions and having captured in excess of 700,000 prisoners in less than three weeks, found the performance of its units, regular as well as reserve, most unsatisfactory. The general staff immediately instituted a massive program to examine the lessons and experiences of the campaign and to pass those lessons on to its divisions through a massive training program. That effort in effect turned the German army into the formidable instrument that broke the back of Allied ground power in spring 1940.[48] The lessons for the Luftwaffe, however, were less clear. In general, tactical execution of missions had been outstanding, and Luftwaffe crews and aircraft had proven generally superior to their opponents. Poland had also been useful in indicating that the Luftwaffe's approach to air war emphasizing the gaining of air superiority was on the right track.[49]

In terms of close air support several important lessons had either been learned or confirmed. On his return from Spain in early summer 1939, Richthofen had been appointed as *Fliegerführer zur besonderen Verwendung* (air commander for special purposes). As such, he received the mission of directing close air support at the critical points of the army's effort. Richthofen then took the first steps toward establishing a capability to support panzer forces in a fluid battle situation. He organized four teams, designated air signal detachments, two of which possessed armored reconnaissance cars and radio equipment, to accompany the mechanized forces with lead units.[50] The strengths of Richthofen's special force suggests the relative priority that the air superiority, interdiction, and close air support missions enjoyed within the Luftwaffe. Richthofen commanded 3 Stuka squadrons, 1 close air support aircraft squadron, 1 Bf 110 squadron, and 1 reconnaissance flight (approximately 114 Stukas, 30 Bf 110s, 20 Hs 123s, and 9 He 126s). Interestingly, nearly 130 Stukas served with other units to support non-close air support missions.[51]

Overall, Richthofen controlled a very small portion of the Luftwaffe's force structure for the close air support mission. His forces supported the Tenth Army's advance and performed yeoman service in the breakthrough of General Walthar von Reichenau's armored forces.[52] Once the mechanized forces had achieved operational freedom, however, the rapid collapse of the Polish defenses and military forces did not provide an ideal laboratory for delineating and defining close air support tactics in fluid battle situations. It took the more complex military operations against Allied forces in the west in spring 1940 to refine close air support doctrine for armored formations.

Nevertheless, much had been learned. By November 1939 General Franz Halder, chief of staff of the army, had signed a new directive establishing a framework within which he hoped army-Luftwaffe cooperation would take place.[53] In particular Halder suggested that the *Kolufts'* foremost responsibility was to coordinate air reconnaissance assigned to the army with those flying reconnaissance missions for the Luftwaffe. Moreover, the army clearly hoped to have the *Kolufts* more directly included and informed as to operational air force intentions and objectives. This resulted from the fact that the *Kolufts* had the most up-to-date reconnaissance information and were in the best position to keep Luftwaffe commanders informed of where the army needed help. Yet the Luftwaffe successfully resisted such a change. As the French campaign showed, Göring had no intention of allowing the *Kolufts*, who were clearly tied to the army, to replace the *Flivos* who were directly within the Luftwaffe's chain of command.

For Luftwaffe formations, the Polish campaign provided a number of lessons. Above all, reported First Air Force (*Luftflotte* 1), communications between ground forces and supporting air force units needed considerable improvement. In high-speed mobile warfare it had proven difficult to keep command authorities informed of movements on the ground.[54] In the case of breakthrough operations (through prepared fortification systems) and defensive lines, First Air Force had found it relatively easy to coordinate with the army as to time, place, target selection and utilization by infantry units for close air support. Interestingly, the Luftwaffe commentators were willing to admit that the material effects of such attacks were not impressive; rather it was the impact on the enemy morale that resulted in significant accomplishments.[55]

First Air Force's "after action" report suggests that the Luftwaffe's impression of events fundamentally differed from the army's. It argued that the *Kolufts* should *not* be more closely included in cooperation between Luftwaffe operational units and the army (outside of the directly assigned close recce squadrons) because they did not possess the necessary ties to Luftwaffe command and control networks to function effectively. It added that the critical element in cooperation would have to be liaison officers possessing good communications, including radio and liaison aircraft, in order to keep up a steady flow of information between the air and ground. In mobile warfare, the "after action" report admitted, a major problem was that the army's command authorities as well as the Luftwaffe's possessed only sketchy information as to the frontline situation. Therefore, advancing troops would have to use smoke and clearly marked recognition devices to indicate who they were (obviously a situation demanding complete air superiority). Fluid situations would also demand security zones within which the Luftwaffe would attack only those ground formations that it could identify with certainty as enemy.[56]

Under First Air Force, the 1st Air Division reported in a similar vein. In particular it singled out the general lack of signal troops within its organization as a major weakness in coordinating the rapid movement forward. It suggested major changes in its TO&E (Table of Organization and Equipment) to repair this deficiency. The importance of good communications had shown up in requests from the army for close air support being forwarded so late that 1st Air Division units could not meet their obligations. Finally, its experience in Poland had suggested that it was going to be difficult to keep air commanders informed of the rapidly changing ground situation.[57] The stress that this divisional "after action" report placed on improved communications is not difficult to understand.

Army "after action" reports from frontline units displayed less satisfaction with the existing level of army-air force cooperation. Some units had nothing to say, suggestive that there had been little cooperation.[58] The 10th Panzer Division, which had played an important part in Guderian's movements, was, however, most dissatisfied. It noted that air reconnaissance had been either late or inaccurate. In one incident the Luftwaffe had reported that fortifications and defensive positions near the Polish town of Lomza were free of Polish troops; 10th Panzer Division's reconnaissance units discovered the Polish defenses occupied by Polish cavalry and other units. More distressing was the fact that the 10th Panzer division's units were constantly machine-gunned and bombed from the air by the Luftwaffe throughout the campaign. One of its units received a particularly graphic demonstration of Luftwaffe effectiveness that left thirteen dead and twenty five badly wounded German troopers. This incident had occurred despite the use of the agreed upon recognition devices by the ground forces.[59]

One of the factors that made the Germans such imposing opponents in both world wars was their ability to absorb and learn from their combat experience at operational and tactical levels. In February 1939 the Luftwaffe High Command had established a tactical experience group as part of its operations division. The new department had the job of examining tactical combat lessons, preparing them in clear, understandable form and then passing them along to frontline units.[60] Under its control, the air staff passed along the tactical and operational lessons of Poland to the flying schools, the operational training units and to those who were preparing for the next campaign: the great offensive against the West. Its ability to absorb the "lessons" of Spain and Poland made the Luftwaffe superior to its opponents in the coming battles.[61]

For the Germans, victory over Poland raised serious strategic problems. Not only had difficulties appeared in the army's performance but the imposition of an Allied blockade had resulted in a ruinous drop in imports with severe implications for the Reich's ability to pursue the war.[62] As a result of the tension between these factors, German strategy in late 1939 led in two separate directions. On one hand, the army pushed for a delay in offensive operations against the West until spring 1940. Hitler, on the other hand, desired an immediate ground offensive to seize the Low Countries and Northern France as a strategic base to strike at Great Britain.[63] The planning for a fall campaign did not aim to replay the Schlieffen Plan or to overthrow Allied military power on the continent.[64] Rather it hoped to achieve limited geographic goals.

Not until January did Hitler finally postpone the western offensive to the spring. By then a new issue had appeared in strategic discussions. Led by Manstein, at that time chief of staff to Army Group A, several officers approached Hitler with an alternative. They suggested that the weight of the offensive be moved from the north to the center to break through French forces deployed along the Meuse River. This breakthrough would be carried out by the bulk of Germany's panzer forces. Once in the open the German armed forces were to race for the English Channel and bottle up Allied forces that had driven into the Low Countries to defend the Dutch and the Belgians against Army Group B's advance. This proposal met considerable opposition at the highest command levels, in particular from the army's chief of staff, Franz Halder. Only in mid-March 1940 did a major war game on the operational prospects of the Manstein plan finally cause the Germans to decide in favor of the new alternative. Even then it was obvious that Hitler and several senior generals felt considerable nervousness. Guderian's memoirs record the following scene at the conclusion of the exercise:

> Hitler asked: "And then what are you going to do?" [after you break through] He was the first person who had thought to ask me this vital question. I replied: "Unless I receive orders to the contrary, I intend on the next day to continue my advance westward. The supreme leadership must decide whether my objective is to be Amiens or Paris. In my opinion the correct course is to drive past Amiens to the English Channel." Hitler nodded and said nothing more. Only General Busch, who commanded the Sixteenth Army on my left, cried out: "Well, I don't think you'll cross the river in the first place!" Hitler, the tension visible in his face, looked at me to see what I would reply. I said: "There's no need for you to do so in any case."[65]

The critical element in Manstein's plan was not how quickly the Germans could get to the Meuse but rather whether they would cross that river and how quickly they could exploit that crossing with their armored mobility. In the wargaming of the Ardennes proposal, it had been clear that mechanized forces would come up on the Meuse by the third or fourth day. Halder had argued that the armor should wait for the infantry divisions to arrive (on the ninth or tenth day) before crossing—precisely what the French expected if the Germans came through the Ardennes.[66] In the end, Halder was persuaded and by April the final plans were set for the mechanized forces to cross the Meuse as soon as they came upon it.

Within the overall plan, the Luftwaffe would play an important role. Its first and most important task was to win air superiority over western Europe by defeating Allied air forces. The achievement of air superiority by a series of major air strikes would place enemy air forces on the defensive and allow the army to execute its operations without serious interference from enemy air attacks. The second, subsidiary task in the early days would be the support of the attack on Holland through airborne drops and, if necessary, bombing attacks in order to eliminate the Dutch as quickly as possible. Third, close air support missions would be laid on during critical moments in the ground battle. The Stuka force, and not the twin-engine bombers, would support ground forces directly, but even the Stukas in the early days of the offensive were to launch strikes against enemy air installations (as part of an overall air superiority strategy). The strength of the two numbered air forces (Second and Third) deployed to support "*Fall Gelb*" (Case Yellow, code name for the offensive) was approximately 1,300 bombers, 860 single-engine fighters, 350 twin-engine fighters, and 380 dive bombers. Thus, the dive bomber force, the only air units specifically trained to support directly the ground advance, numbered less than 15 percent of combat aircraft assigned to the offensive.[67] Second Air Force would support Fedor von Bock's Army Group B's advance with Richthofen's VIII Air Corps (*Fliegerkorps* VIII) for short-range targets and IV Air Corps (*Fliegerkorps* IV) for longer-range objectives. Third Air Force contained I and V Air Corps (*Fliegerkorps* I and V) for longer-range objectives and II Air Corps (*Fliegerkorps* II) to provide close air support at critical moments.[68] Surprisingly, VIII Air Corps (*Fliegerkorps* VIII), the Luftwaffe's most experienced close air support air corps, was not assigned directly to the Ardennes drive—a fact that underlines the importance the Germans placed on destroying the Belgian airfields. The flexibility of air power allowed the Luftwaffe to switch VIII Air Corps to support Third Air Force efforts to expedite the crossing of the Meuse on 14 May.

Right up to the beginning of the offensive the Germans were hard at work attempting to iron out the problems of army-Luftwaffe cooperation. In late April some initial experimental work was conducted to see whether the panzer units could communicate directly with close support aircraft.[69] Nevertheless, the problems of coordinating panzer units with Stukas by means of radio proved too intractable to solve on such short notice.[70] What the Luftwaffe and army had to fall back on were a set of clearly defined bomb lines drawn across the proposed line of operations in France and Belgium.[71] Moreover, during the early days of the campaign a set of carefully delineated ground targets (mostly fortified positions) lying in the path of the advancing panzer forces were selected to receive a pounding from the Luftwaffe, but only *after* its forces had accomplished their air superiority strikes.[72] The communication problems generally reflected the rather sloppy approach that both services took toward supply, and the lack of communications integration is thus not surprising. As late as the Battle of Britain, the Luftwaffe's own fighter and bomber forces were unable to communicate with each other: the former used voice radio, the latter Morse radio transmissions.[73]

On 10 May the offensive began. In a series of major strikes against enemy airfields the Luftwaffe virtually eliminated the Dutch and Belgian air forces. Attacks on British and French air bases in northern France were not as successful but they placed Allied air forces in a defensive posture from which they never fully recovered.[74] Significantly, the Luftwaffe did nothing to interfere with the move to the Dyle in Belgium, which in effect placed Allied troops within a great trap. The Luftwaffe did make a major effort to screen Army Group A's deployment into and through the Ardennes. The first four days of the campaign proceeded as the Germans had hoped. The Luftwaffe had achieved a measure of air superiority over its opponents; Holland was almost out of the war; panzer forces of Army Group A had come up on the Meuse; and the Luftwaffe had already given significant indirect help to troops on the ground by screening the move through the Ardennes from the prying eyes of Allied aircraft.

The key moments in the Battle of France occurred on the Meuse between 13 and 16 May. By the evening of 12 May German armored forces had arrived on the banks of the Meuse, Guderian's XIX Panzer Corps on both sides of Sedan, Reinhardt's XLI to the north of Charleville, and Hoth's XV by Dinant. The decision for an immediate crossing was implicit in the nature of the final Case Yellow plans. By late evening 12 May, operational plans containing mission objectives and specific times for attacking units had been drawn up and passed down the chain of command. Panzer Group Kleist issued its third major order of the campaign at 2330 hours on the 12th. First Panzer Division of Guderian's XIX panzer corps extended and explicated the order at 1200 hours on 13 May in its Divisional Order Number 5. Both orders set the time for the start of the infantry assault across the Meuse as 1600 hours (German time) on the 13th.[75] Both orders made clear the support framework within which the crossings would occur. While the artillery, attached to the panzer divisions, bombarded French positions, II Air Corps would hammer French positions immediately across the Meuse from Panzer Corps XIX. Units attached to VIII Air Corps would support the crossing of Panzer Group Kleist's other panzer corps to the north of Charleville.[76] These attacks would begin at 0800 (German time) and last until 1600 when the crossings began. At that moment bomber forces would shift back away from frontline areas to attack French rear area positions. The Stukas from II Air Corps would support Guderian's crossing and VIII Air Corps' Stukas would aid the crossing above Charleville from 1600 to 1730 hours. After that the Stukas also would shift to interdicting French movement and reinforcements in the rear areas.

If one can describe any combat action as going like clockwork, the Luftwaffe support rendered along the banks of the Meuse came close. A German sergeant with the 1st Panzer Division recalled:

> Three, six, nine, oh, behind still more, and further to the right aircraft, and still more aircraft, a quick look in the binoculars—*Stukas!* . . . Squadron upon squadron rise to a great height, break into line ahead [formation] . . . and there, there the first machines hurtle perpendicularly down, followed by the second, there—ten, twelve aeroplanes are there. Simultaneously like birds of prey, they fall upon their victims and then release their load of bombs on the target. . . . It becomes a regular rain of bombs, that whistle down on Sedan and the bunker positions. Each time the explosion is overwhelming, the noise deafening. Everything becomes blended together, along with the howling sirens of the *Stukas* in their dives, the bombs whistle and crack and burst.[77]

The devastating nature of continuous pounding by Luftwaffe aircraft began the rout that led to a general collapse of French defenses along the Meuse. By evening the rear areas of the French X Corps facing Guderian had become clogged with fleeing troops. In addition, the corps' artillery commander panicked, and pulled his supporting tubes out.[78] The collapse on the Meuse might not have been decisive had not French doctrine been so faulty. Once the Germans had achieved a breakthrough and had gotten their tanks across, the French high command, possessing no strategic reserve, had no chance to plug up the hole. Consequently, the Luftwaffe's employment, in helping the panzer divisions cross the Meuse and break through French positions, played a major role in one of the twentieth century's most decisive strategic victories. Still it is interesting to note that Hoth's panzer corps that crossed the Meuse to the north at Dinant received virtually no air support, while Rommel's account suggests that his 7th Panzer Division (part of Hoth's corps) saw no supporting Luftwaffe aircraft on the 13th and thus crossed the Meuse entirely with its own means. Not until the 15th, when he was rolling towards Philippeville and beyond, did Rommel receive significant air support.[79] The evidence does suggest, however, that the heavy Stuka attacks played a major role in the rapid collapse of the French X Corps, which unhinged the entire Allied position.

The support missions along the Meuse did not represent a revolutionary employment of air power. They were an outgrowth and extension of previous experience, going back to World War I. The effort on 13 May, as with the close air support of the Michael offensive in March 1918, involved the use of aircraft to support infantry who were attacking prepared defensive positions. Consequently, the coordinated army-Luftwaffe support plan, drawn up in outline the night before, targeted known enemy positions. It aimed to achieve a breakthrough of a defensive system about which the Germans already possessed considerable knowledge. It did not involve the coordination and communication difficulties present when the panzers plunged into Allied rear areas at a pace that surprised their own high command almost as much as the rest of the world. Two more points need emphasis: general air superiority played a critical role in the successful intervention of Luftwaffe in the ground battle; and the disastrous failure of Allied efforts to break the logistic links across the Meuse on 15 May. On 12 May five French Curtiss fighters caught 12 Stukas returning from a raid over the Ardennes and shot down all of the German aircraft.[80] Unfortunately, such occurrences were the exception. On the other hand, Allied air attacks on 15 May against the Meuse bridges to isolate German spearheads resulted in such catastrophic losses (the British lost 56 percent of their attacking bombers on that day) that the RAF simply could not resume such attacks with its shattered formations.[81] The contrast between Allied and Luftwaffe aircraft attempting to affect the ground battle was a direct result of German air superiority as well as German doctrine.

The German advance now rolled to the English Channel despite increasing nervousness within the German high command. By Monday, 20 May, elements of Guderian's XIX Panzer Corps had reached beyond Amiens to the Channel coast. The rapidity of the exploitation caused a mad scramble of Allied air units to bases south of the Somme, and Allied air forces played little role in the unfolding events. The Germans were thus able to use their air power uncontested to smash up the Allied rear as well as to render aid in the ground battle. Generally, Luftwaffe missions involved the former rather the latter. In laying on missions for direct ground support, the rapidity of the mechanized advance often rendered such strikes pointless. Bomb lines, flags on forward vehicles, as well as various ground marking devices all were used, but in fact represented no significant advance over First World War procedures.

The victory certainly owed much to air support. The Luftwaffe gained air superiority, suppressed allied reconnaissance missions, blinding Allied armies, helped win the breakthrough battles along the Meuse, and interdicted Allied logistic and reserve movements. The close air support rendered in mobile warfare had been less distinguished, as "after action" reports soon made clear. The most glaring deficiency was the inability of ground units (or for that matter Luftwaffe close reconnaissance units allocated to the army) to communicate with airborne operational units. To put it simply: the Luftwaffe and army still did not possess common radio frequencies.[82] Generally, the Wehrmacht could overcome this handicap where sufficient time existed to coordinate common air-ground operations (such as crossing the Meuse). Such coordination, however, required constant courier flights between Luftwaffe and army headquarters.[83]

When sufficient time did not exist to coordinate or when army units moved so fast as to make coordination difficult, serious problems arose. All the traditional means of identifying ground units from the air were defective. Bomb lines proved difficult to enforce or to coordinate, especially when panzer units began their rampage through northern France. Once mechanized forces reached bomb lines, they faced the disagreeable choice of either stopping, and thus losing what might prove an important opportunity, or of advancing and putting themselves under the threat of air attack by their own air force.[84] Ground recognition devices worked but also presented difficulties. Advancing frontline units did not use their cloth markers to indicate their positions often enough. Where used, they presented German aircraft with a clear mark of where the frontline was, but they were small and hard to see from rapidly moving aircraft. Markers for vehicles and convoys were adequate, although many formations did not mark their columns as prescribed. The use of swastika flags as marking devices was less satisfactory because the red blended in with the color of vehicles, while the white circle was too small for ready identification from higher altitudes.[85]

A major problem, implicit before the offensive began, was the organizational relationship between army and Luftwaffe. Only the rapid French collapse had diminished the impact of an organizational coordination that was less than satisfactory. Manstein's infantry corps reported that while cooperation with close reconnaissance air units had functioned satisfactorily, the assignment of such units shortly before the start of operations had not made sense. To function effectively in combat, the corps "after action" report suggested, close reconnaissance squadrons should be constantly exercised along with the units they would support on the battlefield before operations began.[86] Sixth Army, following the army doctrinal position enunciated after the Polish campaign, argued that it was bad policy to exclude the *Koluft*, who possessed the most up to date reconnaissance knowledge, from input into and coordination of Luftwaffe operations. Thus, it suggested, the *Koluft* should serve not only as the Luftwaffe adviser in all air matters to the chief of staff at army level, but that he should also coordinate the air strikes and supporting missions that lay within the army's sphere of interest. Such a brief would be added to his reconnaissance duties. Under this proposal the *Flivos* from the *Fliegerkorps* would work directly for the *Koluft* and liaison officers from the *Koluft's* staff would be assigned to Luftwaffe units supporting the army.[87]

Such a solution was unsatisfactory to the Luftwaffe and particularly to Hermann Göring. The air force had no intention of allowing the *Kolufts*, who were directly assigned to the army, to interfere in the conduct of Luftwaffe operations. Whatever the outcome of army-air force differences, ambiguities existed throughout the campaign in the functioning of the *Flivo* system—particularly in their ability to communicate up the chain of command. A request from Panzer Group Kleist to VIII

Air Corps met the response that such requests must be passed up to army level and then across to the air corps by the *Flivo* at that level. At the same time Army Group A indicated that such requests could be passed directly to the air corps.[88] Nevertheless, what made the Wehrmacht so effective despite such organizational difficulties was the penchant of the officer corps in both services to take matters into their own hands and to cooperate informally, ignoring whatever command or organizational difficulties existed.

Overall, the Luftwaffe was satisfied with the campaign's operational results and silent on the organizational difficulties that had appeared. As a September 1940 training directive from VIII Air Corps reported, "the earlier combat lessons that had been learned in various campaigns had also once more in the campaign in France and Flanders been confirmed."[89] Close air support of ground troops in critical situations had been of decisive importance. For future close air support operations, VIII Air Corps' chief of staff emphasized the importance of fighter support for dive bomber missions, especially where enemy fighter forces were still operating (undoubtedly a lesson reinforced by the hammering that the Stukas took in the Battle of Britain). Finally, in discussing experiences in rapid mobile operations in France, he stressed that in the swift movements of mobile warfare it would continue to be difficult to gain a clear picture of events on the ground. Therefore, pilots operating over the battle zone must give accurate and immediate reports of what they had seen.[90] Not mentioned in this memorandum was that closer coordination between ground and air signal units might have clarified what was happening on the ground.

The Luftwaffe's contribution to the ground battle and Allied air strategy and capabilities during the same events are an interesting contrast. As the diary of Guderian's panzer corps makes clear, Allied air attacks on advancing German columns early in the campaign caused his troops considerable discomfort. As a result of heavy RAF attacks on the bridges across the Meuse on 14 May, XIX Panzer Corps noted in its diary: "The completion of the military bridge at Donchery had not yet been carried out owing to heavy flanking artillery fire and long bombing attacks on the bridging point. . . . Throughout the day all three divisions have had to endure constant air attacks—especially at the crossing and bridging points."[91] But the losses suffered by the attacking British units were so catastrophic as to render them unfit for further combat. In the largest sense the Allied air forces did not possess a strategy or a doctrine that placed significant emphasis on helping the hard pressed forces on the ground. The RAF's Bomber Command attacked the Ruhr in a series of ill-coordinated and futile attacks; Fighter Command in Britain was removed from the struggle; the RAF's fighters and light bombers in France were assigned tasks all over northern France; and the French had only recently awakened to the threat of air power. Moreover, neither the British nor the French had thought through the problem involved in cooperating extensively with their ground forces. A November 1939 RAF memorandum sums up the British attitude towards close air support: "neither in attack nor in defense should bombers be used on the battlefield itself, save in exceptional circumstances."[92] The result was that the Luftwaffe was able to render significant help to the army, but Allied air forces possessed neither the doctrine nor the desire to give the ground battle the attention called for by its disastrous course.

Between the victory in the west and the onset of operation "Barbarossa" on 22 June 1941 the Wehrmacht was involved in two major campaigns: the Battle of Britain (only the Luftwaffe) and the Balkan campaign of spring 1941. Neither is of particular importance to this study of close air support: the first was almost entirely a Luftwaffe affair, while the latter possessed not only similarities to the French campaign but came so close to coinciding with the invasion of Russia that it had little doctrinal or organizational impact.

Preparations and thinking about the invasion of the Soviet Union began within the army high command on 3 July 1940—even before Hitler turned to that possibility.[93] By December 1940, Hitler had committed the Reich to a massive campaign to destroy the Soviet Union before the onset of winter. To execute such an undertaking the army would carry the bulk of the fighting, and close air support would obviously form an integral part of the effort. The Germans were now on the way towards creation of a coordination system that would be more responsive and that could function with greater effectiveness in a fluid environment.

In the long run several factors impinged on the effectiveness of military preparations. Recognizing that the victory in France had rested largely on the power and combat capabilities of the ten panzer divisions (less than 10 percent of deployed forces), Hitler ordered that the number of such divisions be doubled.[94] At the same time, he recognized the implications of that decision and requested that tank production increase from approximately 120 per month to 800 per month. The army's ordnance department simply rejected the *Führer's* suggestion with the casual and inaccurate observation that such production would ruin the German economy.[95]

Fortunately for her enemies, such attitudes were common in the Wehrmacht after victory over France. As a result, for the next year and a half, the Germans made few substantive efforts to mobilize the economic and productive resources of the continent now at their disposal. Instead, despite the looming campaign against the Soviet Union, German industry continued production at its prewar levels—a result of overconfidence and arrogance rather than of any Blitzkrieg strategy.[96] In a similar vein, Jeschonnek turned away from the frustrations of the Battle of Britain and the failure of the night "strategic" bombing of Britain toward the attack on Russia with the remark: "At last, a proper war!"[97]

The doubling of the panzer divisions coupled with the replacement of obsolete tanks by newer models forced the army to cut the tank TO&E of the armored divisions in half—hardly a recipe for greater sustainability.[98] Similarly, Luftwaffe forces available for the invasion indicate not just a leveling off of German combat power but an actual decrease from the levels attained for the offensive against France: for the French campaign the Luftwaffe had possessed 1,300 long-range bombers, for Russia 775; for France, 380 Stukas, for Russia 310; for France, 860 single-engine fighters, for Russia, 830; for France, 350 twin-engine fighters, for Russia, 90; for France, 300 long-range reconnaissance aircraft, for Russia, 340; for France, 340 tactical reconnaissance, for Russia, 370.[99] Overall, considering aircraft committed to other theaters, the Luftwaffe actually had 200 fewer bombers than at the beginning of May 1940.[100] This decrease resulted from the fact that unlike the army the Luftwaffe had sustained heavy losses from summer 1940 right through to the start of the Russian campaign. Table 6.1 indicates the level of Luftwaffe losses in the bomber and dive bomber forces from May 1940 through May 1941.[101]

TABLE 6.1

LOSSES OF AIRCRAFT ASSIGNED TO UNITS (in percentages)

MAY 1940 TO MAY 1941

	Dive Bombers	Bombers
May 1940	6.8	27.4
June 1940	7.3	12.6
July 1940	2.7	6.0
August 1940	13.7	19.6
September 1940	1.7	18.9
October 1940	1.6	12.3
November 1940	2.4	9.2
December 1940	0.6	5.2
January 1941	2.1	4.8
February 1941	3.0	5.5
March 1941	3.7	8.6
April 1941	10.0	10.6
May 1941	7.2	12.0

In effect, the Wehrmacht was invading the vast spaces of the Soviet Union with little change in the force structure that it had possessed for the invasion of Western Europe.[102] What is particularly surprising, given the German reputation for a high level of military competence, is that few in the high commands of either the army or Luftwaffe found this worrisome. The plans of the Luftwaffe and army resembled the strategic conceptions that had destroyed France the previous year. The Luftwaffe aimed to destroy the Red air force in a massive surprise attack and subsequent operations in the campaign's early days. Similarly the army hoped to surround and liquidate so much of the Red Army in the border areas by its mobile deep penetration thrusts that the Russians would not recover.

In support of the army's operational goal to destroy the Red Army in the border areas, the Luftwaffe was again prepared to render sustained and important help. Close air reconnaissance squadrons were now detailed to each panzer division as well as to the panzer and army corps as had been the case during the previous year. But this improvement had only been achieved by cutting the number of reconnaissance aircraft allocated to each squadron.[103] In addition, the Luftwaffe established air liaison officers down to the panzer division level, in some cases supported by an air signal liaison detachment with a driver and four radio operators. These detachments were assigned to critical areas of the front where mobile operations were taking place. They received armorplated vehicles in order to operate right up with the mobile spearheads.[104] The result was a considerable increase in the Luftwaffe's ability to coordinate air strikes with the army in a mobile environment. Provided that the signal detachments and *Flivos* were up front, the Luftwaffe could now talk to the lead elements of the army's advance on the ground. Ironically, because these detachments reported to their Luftwaffe headquarters by radio, transmissions through their enigma machine were to provide much of the material on army operations that the British intercepted and decrypted in coming years.[105] These reports would play a major role in giving the British a view of what was going on behind German lines in Russia, the Mediterranean, and northwest Europe.[106]

One other major organizational change was made before "Barbarossa." Because several *Fliegerkorps* (in particular II and VIII Air Corps) were responsible for close air support missions and regular interdiction and air superiority strikes, the Luftwaffe established on a permanent basis a *Nahkampfführer* (close air support leader) to control the close air support missions within air corps. That individual was responsible for the movement forward of close air support units, for coordination with the army in the ground battle, and for the communications between the air corps and the panzer groups (later panzer armies).[107]

The administrative change reflected several factors. First, the distances involved in Russia were going to make it likely that the squadrons assigned to air corps would be widely dispersed. It would be difficult for the air corps commander to keep up with the close air support units that would have to move rapidly forward with the army. Moreover, this move may have reflected a desire to meet the army halfway and to provide a means of defusing army efforts to have close air support assets directly under the control of army commanders.

In terms of recognition devices, the Germans had already begun moving away from bomb lines by the start of the invasion. By late summer, in view of the consistently fluid state that had marked operations thus far, the Luftwaffe abandoned their use entirely. In their place came a heavier reliance on marking devices, light signals or pyrotechnics.[108]

In every aspect the Wehrmacht's operations, air as well as ground, lived up to expectations in the campaign's first days. Across the length and breadth of the front, the Luftwaffe caught its opponent by surprise with his aircraft parked in nice neat rows. On the first day IV Air Corps reported destroying 142 enemy aircraft on the ground and only 16 in the air.[109] By noon on the 22d, the Soviets had lost 528 aircraft on the ground and 210 in the air in the western district alone. Along the entire front on 22 June the Red Air Force was to lose 1,200 airplanes in the first eight and one-half hours.[110] Moreover, the rapidly disintegrating situation on the ground forced the Soviets to commit what was left of their air assets in desperate attempts to stem the German tide. Ill-trained, ill-equipped, and badly led Soviet aircrews floundered in impossible formations, from which they were shot out of the skies in huge numbers.[111] Within two weeks of "Barbarossa's" beginning, as two great panzer armies were swinging east towards Smolensk, Halder recorded in his diary:

> On the whole, one can already say that the task of destroying the mass of the Russian army in front of the Divina and Dneper has been fulfilled. I believe the assertion of a captured Russian general to be correct that we can calculate on meeting east of the Divina and Dneper only disjointed forces which alone do not possess the strength to hinder German operations substantially. It is, therefore, truly not claiming too much when I assert that the campaign against Russia has been won in fourteen days.[112]

These enormous successes carried the Germans to Smolensk within a month, placed them three-quarters of the distance to Moscow, and pulled their forces almost to the gates of Leningrad. The Luftwaffe as usual had played a most helpful role. It had gained general air superiority. It supported the army drive directly. Close air support had provided the army's mobile drives into the Soviet Union with enormous help. The new system of mobile liaison had proven particularly helpful in aiding the advance of Second and Third Panzer Groups (armies) toward Minsk and Smolensk. Luftwaffe strikes also caused Soviet military forces irreparable harm in rear areas, and air attacks had broken up numerous Soviet units desperately attempting to escape from German encirclements. Nevertheless, shortly after their arrival at Smolensk the Germans discovered the immensity of their miscalculations. What had worked in France had not worked in Russia. Halder almost despairingly noted in his diary on August 11:

> The whole situation shows more and more clearly that we have underestimated the colossus of Russia—a Russia that had consciously prepared for the coming war with the whole unrestrained power of which a totalitarian state is capable. This conclusion is shown both on the organization as well as the economic levels, in the transportation, and above all, clearly in infantry divisions. We have already identified 360. These divisions are admittedly not armed and equipped in our sense, and tactically they are badly led. But there they are; and when we destroy a dozen, the Russians simply establish another dozen.[113]

Quite simply, the Soviets possessed the strategic depth to absorb the catastrophic defeats on the frontier, while calling up reserves of manpower and production to continue the struggle. The Red air force was in a more difficult situation than the army in that the numbers of aircraft in the early days were harder to replace. Nevertheless, Soviet aircraft production facilities were either outside the range of German offensive operations or were moved as the Germans approached. Thus, a steady and, by the end of the summer, noticeable recovery of the Red air force took place. As that was occurring, the vastness of Russia began to exert its influence. German forces were fanning out across the mouth of a great funnel; and as they did so, they became more and more thinly spread out on the ground. Thus, the ratio of German forces per kilometer, in terms of both troops and firepower, steadily declined. The logistical difficulties of supporting the advance steadily increased.

The same factors impacted on Luftwaffe forces committed to the theater. As early as 5 July, VIII Air Corps reported that fuel was lacking even though it had already severely limited the number of missions flown. Richthofen noted accurately: "Supply is for us the greatest difficulty in this war."[114] Like the army, the Luftwaffe faced almost insurmountable problems in supplying its forward deployed units. The flexibility of air power did allow the Germans to concentrate aircraft rapidly on threatened sectors of the front, but the very extent of that front meant that there were areas where the Luftwaffe could bring nothing to bear. Moreover, declining operationally ready rates due to supply and maintenance problems as well as fuel shortages steadily cut into the Luftwaffe's effectiveness. The Luftwaffe was now spread out across the breadth of the Soviet Union, and while it could still achieve local air superiority wherever it committed sufficient forces, it could not be everywhere. Where it was not, the Red air force could operate with impunity. By late summer, troops on the ground had stopped displaying recognition devices because their visibility was as likely to draw Red air force attacks as they were to warn off the Luftwaffe.[115] Such inclinations were further reinforced by the appearance of Soviet armored ground attack aircraft that proved difficult to shoot down.[116] Unlike the Stuka and earlier German close air support aircraft, these Soviet aircraft had been designed solely for operations over the main battlefield and as such represented a distinct improvement in survivability. Like the T-34, they came as a nasty shock to the Germans.

The spreading out of forces on the ground as well as the rising pressure of Soviet ground attacks led to a drastic increase in army requests for close air support. With scantier resources at its disposal, organizational and administrative improvements in the Luftwaffe could not overcome increasing commitments. As the army's resources spread throughout the theater, it came increasingly to depend on and to demand close air support for offensive operations. As one infantry regimental commander noted: "Tanks in the lead, artillery in the rear, and aircraft overhead—only then will the infantry advance to the attack."[117] By late summer 1941, Hitler as commander in chief of the Wehrmacht ruled that "large-scale offensive operations by the army will only be allowed to commence after the possibility has been insured for extensive support by the Luftwaffe."[118]

Constant combat also exerted great pressure on Luftwaffe capabilities. The VIII Air Corps in twelve days (10 to 21 August) supported I Army Corps in its effort to cut the major Moscow-Leningrad railroad. That action dramatizes the impact of attrition on the Luftwaffe. In this period of supporting just one army corps, VIII Air Corps lost 10.3 per-cent of its aircraft (destroyed or written off as the result of operations) with 54.5 percent damaged but repairable. During this period Rich-thofen's corps also lost 3.9 percent of its flying personnel killed, 5.7 percent wounded, and 2.9 percent listed as missing (for an overall casu-alty rate of 12.5 percent).[119] For the first four months of the Russian campaign the Stuka force was losing or having damaged 20.5 percent of its frontline strength per month. Crew losses over that four month period were nearly 28 percent.[120]

Overall, the Russian campaign drastically increased the attrition of the force structure (a factor that would not end until May 1945). Table 6.2 indicates the extent of the losses in the bomber and dive bomber fleets.[121]

TABLE 6.2

LOSSES OF AIRCRAFT ASSIGNED TO UNITS (in percentages)

	Dive Bombers	Bombers
June 1941	8.1	12.3
July 1941	12.7	18.4
August 1941	8.8	9.7
September 1941	8.5	8.6
October 1941	7.0	10.3
November 1941	Not available	Not available
December 1941	13.8	15.3

Considering the worsening situation on the ground the Luftwaffe now had to commit its regular bomber squadrons to the support of the army's desperate bid to punch through to Moscow and Rostov. Adding to the shortage of aircraft in the east was the removal of much of Second Air Force in November to redress Rommel's critical supply situation in North Africa. Finally in December 1941 the army's advance ground to a halt in front of Moscow in the middle of the Russian winter. When the Soviets went over to the counterattack, the entire German situation in the east trembled on the brink of a complete collapse. In those circumstances the Luftwaffe had no choice but to throw all of its resources into preventing a complete collapse of the army.

The failure of the 1941 campaign in Russia resulted from a fatal overconfidence that had led the Germans to invade with grossly inadequate resources. Close air support undoubtedly contributed to a series of impressive operational victories—Minsk, Smolensk, Kiev, Bryansk, and Vyazma—but even in combination with the army's mobile spearheads, it was not enough to solve the strategic problem: the size and magnitude of resources possessed by the Soviet Union. The system generally functioned more effectively than in France. For the first time one can talk of close air support on a continuous and effective basis in a mobile environment. That is not to say that the system worked flawlessly. In fact, Richthofen's diary is replete with references to a lack of understanding on the part of the army as to the full potential of close air support. Thus, there were times where the army was not fully utilizing the Luftwaffe's capabilities. Moreover, the speed of the army's early advance made it difficult for the divisional and corps level commanders to estimate exactly where their advance had reached. And finally, as Richthofen once noted in the eternal spirit of interservice cooperation, the army was "unteachable."[122] He was undoubtedly giving vent in his diary to frustrations with army commanders who were less than understanding about Luftwaffe difficulties. Nevertheless, the system on the whole worked as well as one could expect given the technology and experiences.

Defeat in the late fall of 1941 represented more than a temporary failure in front of Moscow.[123] In effect it represented the defeat of Germany's effort to gain world hegemony. The Wehrmacht was now deployed deep within the boundaries of European Russia. Its opponent possessed extraordinary recuperative powers in the military as well as the industrial sectors. The Reich's ground forces were spread thinly across the breadth of that theater with limited fire power and almost no reserves, either operational or strategic. Thus, the Luftwaffe had to supply a substantial portion of the missing firepower to bulwark an army that was always in serious straits, even in the summer 1942 offensive. The drive to Stalingrad and the Caucasus represented a most desperate gamble with completely inadequate resources on the ground and in the air. The Luftwaffe had to make up the army's deficiencies in artillery. And as for that firepower support, most of the Luftwaffe was a most inadequate instrument. As has been discussed, only a small proportion of the Luftwaffe, the Stukas, were trained and dedicated to the close air support mission. The Stuka itself was a vulnerable target with no special protection against ground fire. Consequently, while it could drop its ordnance far more accurately than the Soviet ground support aircraft, it was more vulnerable—a factor that showed clearly after 1941 as Soviet forward anti-aircraft defenses began to improve. Ironically, the Germans in 1941 were not even working on a replacement aircraft for the Stuka.

For the future, German ground and air forces in the depths of Russia faced a gloomy situation indeed. Because of the constant recurring crises on the ground, the Luftwaffe had to throw in anything that was available to help the thinly stretched ground forces hold out. Oftentimes the only forces available were the long range bombers and as one Luftwaffe report in December 1942 indicated, such aircraft were unsuitable for the mission. Moreover, most bomber crews did not have the requisite tactical knowledge or training for the close air support mission. Finally, the report emphasized, considering the resources devoted to production of bomber aircraft in terms of engines, size of aircraft, and number of aircrew, such aircraft were entirely cost ineffective compared to aircraft specifically designed for close air support.[124]

As the situation in Russia deteriorated the Luftwaffe found it increasingly difficult to provide the degree of support that the army needed. Specialized antitank, close air support forces were rushed from one sector of the front to another; their operationally ready rate obviously suffered, while the attrition of pilots and aircraft represented a serious drain on their capabilities. One Luftwaffe pilot in an antitank squadron in Russia suggests that his unit lost as many aircraft as the number of tanks that it destroyed—hardly a cost-effective employment of aircraft.[125] Outnumbered in the air, facing heavy anti-aircraft defense on the ground, with its best pilots siphoned off to fight the Allied air forces over the Reich, the Luftwaffe faced an impossible situation on the eastern front—one that neither it nor the army could hope to master.

The picture that emerges from German close air support operations in the first years of World War II is that of a system undergoing considerable development rather than of a clearcut recipe for operational and tactical employment of the Luftwaffe in support of the army. Close air support did not rank among the top missions that Luftwaffe planners foresaw for airpower employment before the war. Even in terms of support for the army, Luftwaffe commanders and planners had a clear preference for the interdiction mission over the close air support mission.[126] There was, moreover, a sense, quite correct as World War II proved, that close air support missions against well defended targets were a costly means of employing air power.

But what the Luftwaffe was willing to recognize, unlike the RAF, was that there would be moments in both offensive and defensive battles on the ground where air power could provide the support on which success turned. Admittedly, the Luftwaffe's approach was tailored to effecting a breakthrough of prepared enemy positions. It was less capable of handling the problems associated with close air support of army formations in a mobile environment. Again, one should not be surprised that this was so. Even the army had considerable conceptual problems as to how mobile operations would evolve, as the various stop orders and command nervousness indicated in the French campaign. Having taken the first steps toward a close air support capability, the Luftwaffe was able to refine that capability for the Russian campaign. Nevertheless, whatever contribution the Luftwaffe made to the advance of ground forces, no matter what operational brilliance the army might show in executing its orders to destroy the Red Army, the Wehrmacht could not overcome the handicaps of a high command (and not just Hitler) drunk with victory, of a logistical system that functioned badly because of command negligence, and of the vastness of Soviet resources and space.

As with so much of German military history in the twentieth century, one comes away with a sense that in the operational sphere the army and Luftwaffe represented military organizations that willingly grappled with difficult problems in a realistic, rational fashion. They could and did learn lessons from combat experience, and then applied those lessons in preparing for the next battles. German military effectiveness also rested on a spirit of wholehearted cooperation at the lower and intermediate levels both within and between services. Consequently, operations tended to run smoothly and with little bickering over roles and missions between the services. Most German officers seem to have felt that the lives of aircrews and ground troops and the successful completion of military operations were more important than the narrow concerns of their own service. The strategic and political spheres were, of course, another matter.

NOTES

1.   The basis of this statement rests on considerable research done at the Militärarchiv and the Militärgeschichtliches Forschungsamt in Freiburg in June 1984.

2.   General der Kavallerie von Hoeppner, *Deutschlands Krieg in der Luft* (Leipzig, 1921), pp. 114, 117-18, 149-50.

3.   See Timothy Lupfer, *The Dynamics of Doctrine: The Changes in German Tactical Doctrine During the First World War* (Leavenworth, 1982).

4.   Rittermeister Manfred Frhr. von Richthofen, *Sein militärisches Vermächtnis, Das Testament, Die Erfahrungen im Luftkampf*, edited by the Luftwaffe historical section (Berlin, 1938); copy provided by General Hans Asmus, Baden-Baden, 23 June 1984.

5.   Hoeppner, *Deutschlands Krieg in der Luft*, pp. 149-50.

6. Reichsarchiv, *Der Weltkrieg, 1914 bis 1918*, vol. 14, *Die Kriegführung an der Westfront im Jahre 1918* (Berlin, 1944), pp. 720-21.

7.   For a discussion of how this process worked, see Williamson Murray, *The Change in the European Balance of Power, 1938-1939, The Path to Ruin* (Princeton, 1984).

8.   Dennis Richards, *The Royal Air Force, 1939-1945*, vol. I (London, 1953), p. 29.

9.   For the first statement of this revisionist position, see Klaus Maier's discussion in Klaus A. Maier, Horst Rohde, Bernd Stegemann, and Hans Umbreit, *Das Deutsche Reich und der Zweite Weltkrieg*, vol. II, *Die Errichtung der Hegemonie auf dem europäischen Kontinent* (Stuttgart, 1979). For a more detailed discussion of this point in English, see Williamson Murray, *Luftwaffe* (Baltimore, 1985), and "The Luftwaffe before the Second World War: A Mission, A Strategy?," *Journal of Strategic Studies* (September 1981).

10.   Edward L. Homze, *Arming the Luftwaffe, The Reich Air Ministry and the German Aircraft Industry, 1919-1939* (Lincoln, 1976), pp. 40-41.

11.   For the general unwillingness of the senior German leadership to think on the strategic level see Williamson Murray, "JCS Reform, A German Example?", *JCS Reform*, edited by Steven Ross (Newport, 1985).

12.   I am indebted to Oberstleutnant Dr. Klaus A. Maier for drawing my attention to Wever's lecture. See: "Vortrag des Generalmajors Wever bei Eröffnung der Luftkriegsakademie und Lufttechnischen Akademie in Berlin-Gatow am 1. November 1935," *Die Luftwaffe* (1936).

13.   Again I am indebted to Oberstleutnant Klaus Maier for providing me a copy. See "Die Luftkriegführung," (Berlin, 1935).

14.   Ibid., p. 21.

15.   Ibid., pp. 125-32.

16.   For the difficulties involved in developing the He 177, see Edward R. Homze, "The Luftwaffe's Failure to Develop a Heavy Bomber Before World War II," *Aerospace Historian* (March 1977).

17.   Wilhelm Deist, Manfred Messerschmidt, Hans-Erich Volkmann, Wolfram Wette, *Das Deutsche Reich und der Zweite Weltkrieg*, vol. I (Stuttgart, 1979), pp. 480-81.

18.   There were some in the air ministry in 1933 who went so far as to argue that the *whole* emphasis in Germany's air rearmament should be on the creation of a four engine strategic bombing force. The technical capabilities of Germany's aircraft industry at the time made such a proposal completely unrealistic. See Bernard Heimann and Joachim Schunke, "Eine geheime Denkschrift zur Luftkriegskonzeption Hitler-Deutschlands vom Mai 1933," *Zeitschrift für Militärgeschichte* 3 (1964), pp. 72-86.

19.   For the inadequacies of the theories of a "Blitzkrieg strategy" in the prewar period, see Murray, *The Change in the European Balance of Power, 1939-1939*, chapter I.

20.   Conversation with General Hans W. Asmus, Baden Baden, 23 June 1984. In fairness to Guderian, he did at least see that the aircraft could play an important role in furthering the exploitation drive of tank forces. See Heinz Guderian, *Schnelle Truppen Einst und Jetzt* (Berlin, 1936), p. 236.

21.   Hitler understood that it was to Germany's advantage that the war in Spain serve as a distraction from the growing threat posed by German rearmament and that outside of its ability to distract the European powers Spain had little strategic significance for the immediate future of the Reich. See Gerhard Weinberg, *The Foreign Policy of Hitler's Germany, 1933-1936*, vol. I (Chicago, 1970), p. 298.

22.   Matthew Cooper, *The German Air Force, 1933-1945, An Anatomy of Failure* (New York, 1981), p. 59.

23.   Ibid., p. 59.

24.   Derek Wood and Derek Dempster, *The Narrow Margin* (London, 1961), pp. 49-50. The finger four formation involved the combination of two ship formations in which the second pilot covered the tail of the lead aircraft and all four pilots were responsible for predetermined areas of the sky to insure that the formation was not taken by surprise.

25.   Conversation with General Asmus, Baden Baden, 7 and 8 November 1980 and letter from General Asmus, 6 February 1981.

26.   Air Ministry, *The Rise and Fall of the German Air Force, 1933-1945* (London, 1948), pp. 16-17. For the considerable difficulties that Richthofen faced and had to overcome, see the entries in his Tagebuch for March and April 1937 that refer to the shortcomings and misunderstandings that stood in the way of establishing even a relatively primitive system.

27.   Air Ministry, *The Rise and Fall of the German Air Force*, pp. 16-17; and "Lehren aus dem Feldzug in Spanien, Einsatz von Schlachtfliegern," aus einer Studie der 8. Abt. des Generalstabes aus dem Jahre 1944; Hans Hennig Freiherr von Beust, "Die deutsche Luftwaffe im spanischen Krieg," 2.10.56, p. 162, AFSHRC:K 113.302.

28.   Conversation with General Asmus on 23 June 1984, Baden Baden, Federal Republic of Germany. General Asmus indicated that the flags were carried by advancing infantry as well as spread out as recognition devices.

29.   Bundesarchiv/Militärarchiv (BA/MA) RL/57, Auswertung "Rügen", Anlage 2 zu Lw.Gr.Kdo.3, Nr. 7179/38g.Kdos, Heft 2, a)Führung Abschnitt IV bis VI, p. 50.

30.   Conversation with General Asmus on 23 June 1984, Baden Baden, Federal Republic of Germany.

31.   For a fuller discussion of Germany's economic difficulties in the late 1930s and the constraints that this placed on German rearmament, see Murray, *The Change in the European Balance of Power, 1938-1939*, chapter I.

32.   Air Historical Branch, Air Ministry, vol. VII, Translations: "Luftwaffe Strength and Serviceability Statistics," G302694/AR/9/51/50.

33.   BA/MA, Za 3/109, "Studie zum Schlachtfliegereinsatz," Obersting Cornelius and Major Brüker, 8.Abt., XII. 1944.

34.   BA/MA, RL 7/155, Lw. Gruppenkommando 3, Abt. Ic, 23 Mai 1938, "Bericht über die Reise für Führer oberer Dienststellen, Thüringen 15-22.5.38," Section IV Zusammenarbeit mit der Luftwaffe, pp. 9-10.

35.   BA/MA RL 7/158, Luftflottenkommando 3, Führ. Abt/Ia op/Nr.93/39, 17.5.39, "Bericht über die Heeresgeneralstabsreise 1939," Anlage 1, "Beitrag für Schlussbesprechung (Luftwaffe)," p. 2.

36.   BA/MA, RL7/159 Verlauf der Generalstabsreise Luftwaffe 1939, Lft. Kdo 3, Fhr.Abt. Nr 2778/39, pp. 6-17.

37.   BA/MA, Lw 106/15, Der Einsatz der dem Heer taktisch unterstellten Verbände der Luftwaffe.

38.   Unpublished paper by Hauptmann v. Gyldenfeldt, "Die Beurteilung des Luftwaffeneinsatzes während des Norwegen- und Frankreich-Feldzuges 1940 durch die Heeresführung," Führungsakademie der Bundeswehr, Hamburg 1970; copy made available to me by Dr. Horst Boog, Militärgeschichtliches Forschungsamt, Freiburg, Federal Republic of Germany.

39.   Conversation with General Asmus, Baden-Baden, Federal Republic of Germany, 24 June 1984. General Asmus stated that the Luftwaffe close reconnaissance squadron commander worked directly for the army corps commander and had virtually nothing to do with the division commanders. Rommel's comments on the campaign hardly suggest the assignment of a close reconnaissance squadron to his panzer division (see Erwin Rommel, The Rommel Papers, ed. B. H. Liddell Hart [London, 1953], chapter I). But BA/MA RH 19 III/152, "Taktische Erfahrungen im Westfeldzug," Oberkommando des Heeres, GenStdH/ Ausb.Abt. (Ia), Nr.2400/40, 20.11.40 does, however, suggest that close reconnaissance squadrons were at times assigned directly to panzer divisions in special circumstances.

40.   Gyldenfeldt, "Die Beurteilung des Luftwaffeneinsatzes während des Norwegen- und Frankreich-Feldzuges 1940 durch die Heeresführung."

41.   Conversation with General Asmus, Baden Baden, Federal Republic of Germany, 24 June 1984.

42.   Ibid.

43.   Ibid.

44.   General der Flieger Paul Deichmann, German Air Force Operations in Support of the Army, USAF Historical Study, No. 163 (Montgomery, 1962), p. 153.

45.   I am indebted to Dominic Graham for this point and quotation. His trenchant, carefully thought out comments on an earlier draft of this manuscript were enormously helpful.

46.   "The Luftwaffe in Poland," a study produced by the German Historical Branch (8. Abteilung), 11.7.44, AHB, Translation No. VII/33.

47.   For an account of German Stukas along the Bzura, see Rolf Elbe, Die Schlacht an der Bzura im September 1939 aus deutschen und polnischer Sicht (Freiburg, 1975).

48.   For a fuller discussion of this process, see Williamson Murray, "The German Response to Victory in Poland: A Case Study in Professionalism," Armed Forces and Society (Winter 1981).

49. For a fuller discussion of Germany's air superiority strategy see Williamson Murray, "Poland and the West" in the Air Superiority sister volume to be
published by the Office of Air Force History.

50. Deichmann, *German Air Force Operations in Support of the Army*, p. 131.
Deichmann implies they accompanied panzer divisions in the Polish campaign.
It is more likely that they were assigned to corps level as the campaign evolved.

51. BA/MA, Lw 106/14, "Die Unterstützung des Heeres durch die deutsche
Luftwaffe im zweiten Weltkrieg" (Versuchsweise Aufstellung einer
Staffgliederung mit Erläuterungen) von. Gen.d.Fl. Paul Deichmann, pp. 40-52.

52. BA/MA, RL7/340 Luftflotte 4. Divisionbefehle, Flieg. Div. 2., Ia Nr. 70/39,
1.9.39, Befehl für den Einsatz am 2.9.39.

53. BA/MA, H35/88, Oberkommando des Heeres, GenStdH/Ausb.Abt. (Ia)
Nr. 750/39, Richtlinien für die Zusammenarbeit Heer-Luftwaffe auf Grund der
Erfahrungen im polnischen Feldzuge.

54. BA/MA, RL7/2, Abschrift, Erfahrungsbericht der Luftflotte 1 über den
polnischen Feldzug, p. 1.

55. Ibid., pp. 9-10.

56. Ibid., pp. 11-12.

57. BA/MA, RL 7/2 Anlage zu Fliegerdivision 1, Br.B.Nr. 3185/39, "Vor-
läufiger Erfahrungsbericht über den Einsatz während des poln. Feldzuges."

58. 20. Division, Abt.Ia Nr. 510/39, Betr.: Erfahrungen, 4.10.1939, National
Archives and Records Service NARS, T-314/614/000656.

59. 10. Panzer Division, Abt.Ia Nr. 26/39, Erfahrungsbericht, 3.10.1939,
NARS, T-314/614/000632.

60. Deichmann, *German Air Force Operations in Support of the Army*, p. 56.

61. Because most of the Luftwaffe's records were destroyed at the end of
the Second World War, it is difficult to evaluate how this process worked. The
records of the army are more complete and thus one can more fully evaluate
how vigorously combat experience was worked into the training programs and
preparations for the next campaign. See Murray, "The German Response to
Victory in Poland: A Case Study in Professionalism," chapter 9.

62. For a fuller discussion of the strategic and economic factors involved in
the first six months of the war, see Murray, *The Change in the European Balance
of Power, 1938-1939*, chapter 10.

63. For a fuller discussion see Murray, *Luftwaffe*, chapter 2.

64. The Schlieffen Plan had been designed before World War I by the Germans to overwhelm France on the outbreak of war by launching a great out-
flanking move through Belgium. It failed on the Marne in September 1914.

65. Heinz Guderian, *Panzer Leader* (New York, 1953), p. 92.

66.   Telford Taylor, *March of Conquest* (New York, 1958), pp. 168-71.

67.   Air Ministry, *The Rise and Fall of the German Air Force*, p. 66.

68.   Taylor, *The March of Conquest*, p. 183.

69.   1. Panzer Division, Ia Nr. 232/40, 24. April 1940, "Zusammenarbeit Panzer-Stuka," NARS, T-314/615/00393.

70.   Generalkommando XIX.A.K., Ia/NaKa Nr 362/40, 6.5.1940, "Zusammenarbeit Panzer-Stuka," NARS, T-314/615/00396.

71.   Nahkampfführer II, Ia Nr 58/40, "Sicherheit der eigenen Truppe vor eigenen Luftangriffen," 4.4.40, NARS, T-314/615/00358.

72.   Nahkampfführer II, IaNr 145/40, 30.4.40, Befehl Nr..2, Befehl für die Kampfführung des Nahkampfführers II am A-Tag ausser dem ersten Einsatz gegen die französische Luftwaffe, NARS, T-314/615/00377.

73.   Francis K. Mason, *Battle over Britain* (London, 1969), p. 237.

74.   BA/MA, Lw 106/15, "Überblick über den Einsatz der Luftwaffe bei den Operationen in den Niederlanden," in Belgien, und Nordfrankreich, D.Ob.d.Lftw. Führungsstab Ic Nr. 10641/40, 3 Juni 1940.

75.   For the 1st Panzer Division order see BA/MA Lw. 106/15, "Befehl der 1. Panzerdivision zum Übergang über die Maas am 13.5.40 mit gemeinsamem Feuerplan für Luftwaffe und Heer," Ia, 13.5.40, 12 Uhr. For the Panzer Group see v. Gyldenfeldt, "Die Beurteilung des Luftwaffeneinsatzes während des Norwegen- und Frankreich-Feldzuges durch die Heeres Führung," Anlage 9.

76.   BA/MA RL 8/45 Gen. Kdo. VIII. Fliegerkorps. "Einsatz im Feldzug gegen Frankreich (Fragment eines Tagliches-Abschrift)."

77.   Quoted in Alistair Horne, *To Lose a Battle, France 1940* (Boston, 1969), p. 289.

78.   Jeffery A. Gunsburg, *Divided and Conquered, The French High Command and the Defeat in the West, 1940* (Westport, Conn., 1979), p. 190.

79.   Rommel, *The Rommel Papers*, pp. 7-15. Rommel's account on how he was informed of air support for the 15th by a Luftwaffe major, clearly not familiar to him, suggests that the Flivos were not yet assigned below corps level.

80.   Horne, *To Lose a Battle*, p. 253.

81.   Calculation based on mission reports in Major L. F. Ellis, *The War in France and Flanders* (London, 1953), pp. 55-56.

82.   Hans-Adolf Jacobsen, *Dokumente zum Westfeldzug 1940*, (Göttingen, 1960), p. 79.

83.   See BA/MA, RL 8/45, Fliegerkorps VIII, War Diary, pp. 5-6, 15-6, 5.40 for the constant comings and goings of commanders and liaison officers.

84.   BA/MA, Lw 106/4, "Die Unterstützung des Heeres durch die deutsche Lw. im zweiten Weltkrieg (Versuchsweise Aufstellung einer Stoffgliederung

mit Erläuterungen)" von Gen.d.Fl. Paul Deichmann.

85.   BA/MA, RH12-5/v.246, Armee-Oberkommando 6, Ia Az 2 Nr. 3104/40, 10 August 1940, Betr.: Erfahrungsbericht.

86.   v.Gyldenfeldt, "Die Beurteilung des Luftwaffeneinsatzes während des Norwegen- und Frankreich-Feldzuges 1940 durch die Heeresführung," p.11.

87.   BA/MA, RH 12-5/v.246, Armee-Oberkommando 6, IaAz 2 Na. 3104/40, 10 August 1940, Betr: Erfahrungsbericht.

88.   v.Gyldenfeldt, "Die Beurteilung des Luftwaffeneinsatzes während des Norwegen- und Frankreich-Feldzuges 1940 durch die Heeresführung," p. 12.

89.   BA/MA, RL 8/250, Generalkommando des VIII. Fliegerkorps, Abt.Ia, 21.9.1940, "Richtlinien für Ausbildung und Einsatz der zur Unterstützung des Heeres eingesetzten Fliegerverbände des VIII. Fliegerkorps."

90.   Ibid., p. 11.

91.   Quoted in Ellis, The War in France and Flanders, p. 56.

92.   Public Record Office, CAB 21/903, 18.11.39, "Bomber Support for the Army," memorandum by the air staff; see also the letter from Admiral Lord Chatfield to Prime Minister Chamberlain, 15.11.39. on the air force arguments against training special units to cooperate with the army.

93.   Horst Boog, Jürgen Förster, Joachim Hoffmann, Ernst Klink, Rolf-Dieter Mller, Gerd R. Ueberschär, Das Deutsche Reich und der Zweite Weltkrieg, vol. IV, Der Angriff auf die Sowjetunion (Stuttgart, 1983), p. 9.

94.   Ibid., pp. 168-89.

95.   Guderian, Panzer Leader, p. 114.

96.   For a full discussion see Murray, Luftwaffe, pp. 92-106.

97.   David Irving, The Rise and Fall of the Luftwaffe (Boston, 1973), p. 123.

98.   For the full defects in German equipment preparations for Barbarossa see: Boog, et al., Das Deutsche Reich und der Zweite Weltkrieg, pp. 186-87.

99.   Air Ministry, The Rise and Fall of the German Air Force, pp. 66, 165.

100.   Comparative figures available in "Luftwaffe Strength and Serviceability Tables, August 1938-April 1945," AHB, Translation VII/107.

101.   BA/MA, RL 2 III/1025, Genst.6.Abt. (III A), Front-Flugzeugverluste.

102.   The German pre-invasion war games had indicated most clearly the implication of Russia's vastness. See George Blau, The German Campaign in Russia - Planning and Operations (1940-1942) (Washington, 1955), p. 20.

103.   Deichmann, German Air Force Operations in Support of the Army, p. 70.

104.   Ibid., p. 132.

105.   See Ralph Bennett, Ultra in the West (New York, 1980).

106.   One of the ironies of the "alliance" with the Soviet Union is the fact that the British garnered more information about what was going in military opera-

tions in the east from enigma decypts than from our Soviet "allies."

107.   BA/MA, RH 27-18/14, Oberkommando des Heeres, Gen StdH/Ausb.Abt. (Ia), H.Qu.Okh., 26.5.41, Nr. 1161/41, "Taktisches Merkblatt für die Führung von Nahkampf-Verbänden.

108.   Deichmann, *German Air Force Operations in Support of the Army*, pp. 134-35.

109.   BA/MA, RL 8/31, Generalkommando des IV. Fliegerkorps, Abt IC, "Lagebericht v. 22.6.41."

110.   John Erickson, *The Road to Stalingrad* (New York, 1975), pp. 118-19.

111.   Albert Kesselring, *A Soldier's Record* (New York, 1953), p. 90.

112.   Franz Halder, *Kriegstagebuch*, Vol. III, (Stuttgart, 1964), p. 38.

113.   Ibid., p. 170.

114.   BA/MA, RL 8/49, Russland-Feldzug 1941, VIII. Fliegerkorps.

115.   Boog, *Das Deutsche Reich und der Zweite Weltkrieg*, vol. IV., p. 658.

116.   Ibid., p. 661.

117.   Deichmann, *German Air Force Operations in Support of the Army*, p. 126.

118.   Ibid., p. 126.

119.   BA/MA RL 8/47, Generalkommando I.A.K., Abt.Ia 545/41, 16.9.41, "Einsatz des Fliegerkorps VIII vom 10.-21.8.41," Anlage 1, Tätigkeit des VIII. Fliegerkorps bei der Unterstützung des Durchbruchs des I.A.K. bis zur Eisenbahnlinie Leningrad-Moskau vom 10.8.41-21.8.31.

120.   Based on figures in BA/MA, RL 2 III/715, Gen. Qu.6 Abt. (I), "Übersicht über Soll, Istbestand, Verluste und Reserven der fliegenden Verbände," 1.11.41.

121.   BA/MA RL 2III/1025, Genst. 6.Abt. (IIIA), Front-Flugzeugverluste.

122.   BA/MA, RL 8/47, VIII. Fliegerkorps, zusammengestellt von H. W. Deichmann . . . von Aufzeichnungen, Umfragen, und Tagebuch G.F.Frhr. von Richthofen. See entries for 25.6., 26.6.41 in particular.

123.   For a full examination of the reasons for the German failure see: Klaus Rheinhardt, *Die Wende vor Moskau* (Stuttgart, 1977).

124.   BA/MA, RL 7/8, Luftwaffenkommando 1, Nr. 5585, 21 Dezember 1942, An den Oberbefehlshaber der Luftwaffe, Führungsstab Ia.

125.   Murray, *Luftwaffe*, p. 236.

126.   Deichmann, *German Air Force Operations in Support of the Army*.

CHAPTER
SEVEN

THE LUFTWAFFE
AGAINST POLAND AND THE WEST

When World War II began on 1 September 1939, the generally held assumption among airmen was that "strategic" bombing would be the dominant mode in which air forces would fight the coming war. The Germans held assumptions similar to beliefs that were current in the U.S. Army Air Force and the RAF. But they were also open to a wider strategic view on the employment of air power: that the destruction of the enemy air force and the achievement of air superiority and support for the army's efforts on the ground (particularly interdiction) were equally worthwhile tasks. As a result the Luftwaffe was the best prepared of all the world's air forces in 1939 to fight a realistic campaign to support overall military objectives. This chapter lays out the general framework within which the Luftwaffe approached the problem of air superiority before and during the first campaigns of the Second World War. It aims to help achieve an understanding of the strengths as well as the weaknesses of the German approach to air superiority and how those strengths and weaknesses contributed to the campaigns of 1939 and 1940.

When World War I ended in November 1918 not much seemed clear about the role of air power in modern war except that it represented a dimension that no major nation could safely ignore. Though the full employment potential of aircraft was somewhat uncertain, the conflicting powers had employed aircraft in virtually all the roles in which they have appeared through to the present; air superiority, "strategic" bombing, close air support, interdiction, and photo reconnaissance all played their part (only air transport did not receive significant attention). Nevertheless, the lack of clarity over the lessons of World War I led many interwar theorists to emphasize the theoretical and to ignore the practical realities of air power.

One lesson should not have been ambiguous; the fundamental principle on which all World War I air operations rested was the need for air superiority. Without that basic attribute, photo reconnaissance aircraft did not return with intelligence; tactical bombers on close air support or interdiction strikes suffered shattering casualties; and "strategic" bombers suffered prohibitive losses that soon ended bombing campaigns. On the other hand, bombers and reconnaissance aircraft, sheltered by air superiority, could carry out their missions without prohibitive losses. The achievement of air superiority, however, posed a difficult and costly challenge. Even when air forces gained local superiority, whether through quantitative or qualitative advantages, that superiority usually proved transitory; the enemy could master numerical inferiority by reinforcing contested sectors. In the case of qualitative inferiority, he could redress technological imbalances by advances of his own. This resulted in an air war with shifting balances and heavy casualties. Ironically, the lessons of the importance of air superiority and the difficulties inherent in achieving it did not strike a responsive chord among interwar air power theorists.

The seeming paucity of "lessons" on other aspects of aircraft employment failed to inhibit evolution of theories arguing that the aircraft would be *the* dominant weapon of the next war. Two major trends in such thinking evolved: the ancestors of the modern schools of counterforce and countervalue nuclear strategies. British theorists placed primary emphasis on direct attacks on enemy population centers (particularly the working class), while American theorists stressed the vulnerability of enemy economic systems to precision bombing attacks directed at critical points in the industrial structure. As the future Air Marshall Sir John Slessor suggested in 1936, a nation could gain and maintain air superiority only through a "resolute bombing offensive" against enemy cities and industries.[1] A more general, but certainly representative, discussion on air power in a future conflict appeared in an RAF Air Staff memorandum of 1924 that argued that air forces:

> can either bomb military objectives in populated areas from the beginning of the war, with the objective of obtaining a decision by moral[e] effect which such attacks will produce, and by the serious dislocation of the normal life of the country, or, alternatively, they can be used in the first instance to attack enemy aerodromes with a view to gaining some measure of air superiority and, when this has been gained, can be changed over to the direct attack on the nation. The latter alternative is the method which the lessons of military history seem to recommend, but the Air Staff are convinced that the former is the correct one.[2]

Thus, a major theme in prewar thinking was that the traditional strategic factors would not bind air power. In other words, the aircraft had negated the principles of war.[3]

While a few airmen like "Billy" Mitchell addressed the problems of gaining and maintaining air superiority, other interwar theorists denigrated not only defensive air war but also strikes against enemy air power. In fact, many airmen regarded such strategies as a waste of effort. There was, of course, evidence supporting the belief that "the bomber will always get through;" most notably bombers evolved more quickly than fighters in the 1930s and consequently, it proved difficult to envision a successful interception of enemy air fleets. Nevertheless, a minimization of the possibility of fighters or bombers attacking enemy air forces or air bases also reflected ideological beliefs that "strategic" bombing was the only proper employment for aircraft. As Sir Hugh Trenchard somewhat crudely expressed it while discussing a possible air war with France:

I would like to make this point again. I feel that although there would be an outcry, the French would probably squeal before we did [in an air war between France and Great Britain]. That was really the first thing. The nation that would stand being bombed longest would win in the end.[4]

The general historical view has tended to place the Luftwaffe outside the mainstream of the interwar air power theories. It suggests that the many German Army officers transferring to the new service in 1933 brought with them only narrow, land-war oriented concerns.[5] Thus, supposedly, the Luftwaffe remained closely tied to the army with neither interest nor understanding of "strategic" bombing. Reinforcing this view has been a historical construct, the so-called "Blitzkrieg" strategy, that argues along the following lines; the Nazi leadership, faced with certain economic and political preconceptions, evolved a grand strategy suited to Germany's peculiar needs. This strategy, the argument continues, did not include rearmament in depth, but created an elite panzer force, supported by the Luftwaffe, to fight short, quick, campaigns to avoid a long drawn out war.[6]

Unfortunately, this generally accepted view now appears erroneous. German grand strategy and its air component did not follow an obvious or consistent path. Rather there existed at the highest level an almost complete lack of strategic planning. Admittedly, Hitler possessed a clear sense of his long range goals: to destroy the European constellation of power and to establish in its place a Europe under German control entirely free of Jews and "Jewish-influences." While the destruction of the diplomatic balance from 1933 to 1936 proved surprisingly easy,[7] a combination of massive rearmament, foreign exchange difficulties, and other problems caused political and economic crises after 1936 that the outward thrust of German policy has obscured.[8] Those difficulties prevented the Germans, including Hitler, from framing a coherent national defense policy. The Führer, of course, maintained a firm sense of the ultimate objective toward which he was driving both state and military. Nevertheless, he worried little over the means available, while his generals, with the possible exception of the army's chief of the general staff, General Ludwig Beck, never worried overly about strategic questions. One comes away from the sources with a sense of an extraordinarily chaotic defense policy. What is especially surprising, given the current reputation that Germans enjoy in military affairs, was the lack of centralized control or even generally accepted goals among the military and the cavalier disdain that the services showed towards economic realities throughout the rearmament process.[9]

Within a sea of contending forces, the Luftwaffe found its interests well-protected by Hermann Göring's position in the Nazi political structure. But in its formative years others helped chart the Luftwaffe's course. The two most important individuals were the State Secretary, Milch, and the first chief of staff, Wever. Early on, the Luftwaffe's leaders considered an all "strategic" bomber force structure[10]—a theme that struck a responsive cord in the Luftwaffe throughout the prewar period.[11] But Milch and Göring rejected the proposal, not because "strategic" bombing was foreign to their *Weltanschauung,* but rather because Germany's industrial, technological, and geographic situation made a "strategic" bombing force unrealistic for the immediate future.[12]

Wever largely cast the Luftwaffe's strategic framework in the 1930s. While Milch handled the economic and administrative tasks of creating the new military service and Göring took care of politics, the chief of staff established the intellectual and strategic patterns within which the Luftwaffe grew. Despite a lack of aircraft experience, he had received his appointment as the Luftwaffe's first chief of staff.[13] In the short period before his untimely death in 1936, Wever exercised an extraordinary influence over the Luftwaffe's basic doctrine.

Two documents spell out his thinking on the question of air power: a speech to the *Luftkriegsakademie* (Air War College) in November 1935, and the Luftwaffe's basic doctrinal manual, *"Die Luftkriegführung"* (The Conduct of Air War) published in late 1935.[14] Wever argued for a more broadly based approach to air power than did most other theorists in this period. He was never an unabashed champion of "strategic" bombing, but rather suggested that a variety of factors would determine the Luftwaffe's role in any future war: the overall strategic situation, the weather, national objectives, and the nature of enemy forces among others. In particular, one could not easily separate the struggle against enemy air forces from the support that the Luftwaffe would have to provide the army and navy. Even though its flexibility of employment gave it advantages over ground and naval forces, its primary opponent would be enemy armed forces, particularly the air force. Wever argued that gaining air superiority, whether local or general, represented a most difficult goal. Changing technology, new aircraft types, and re-placement by new production and freshly trained crews would allow an enemy air force to return and fight again.[15] Air superiority would demand an unremitting commitment. Nevertheless, like most interwar air power theorists, he believed that the bomber would be the decisive weapon of aerial warfare.[16] While one could and should rely on active as well as passive defensive measures, the best method of defeating the enemy in the air, he contended, was to strike at the basis of his air power: in particular at his bomber fleet on the ground and at the indus-trial support that allowed the enemy to make good his losses.[17] The Luftwaffe's doctrinal manual made clear that the enemy's *air force* was the primary target at the beginning of war:

> One must attack the enemy's air force from the beginning of war.
> Its defeat will weaken the enemy's armed forces, while protecting
> one's own air force to carry out other missions important to the
> war effort. The struggle aims pre-eminently at the enemy's
> bomber strength. First of all mobile units must be destroyed.
> Surprise strikes of one's own bombers at the beginning of war
> can succeed in hitting the enemy's bombing power at peace time
> bases.[18]

Between Wever's death and the outbreak of war, the Luftwaffe developed into a formidable instrument. That expansion from a non-existent force in 1933 to the most powerful air force in the world in 1939 with over 4,161 aircraft (including 1,179 fighters and 1,180 bombers)[19] imposed a considerable strain on the national economy. The Germans not only faced the task of acquiring the technical and operational exper-tise necessary for such a force,[20] but within the space of six years they virtually had to replace it with a new generation of aircraft.[21]

The prewar development of German operational concepts was con-siderably influenced not only by theory but by Germany's exposed strategic situation, the megalomaniacal goals of her leader, and actual combat experience from Spain. War games conducted as early as 1934 suggested that direct attacks on an enemy's air force and bases would not entirely eliminate his bombing capability. Therefore, the Germans concluded that they needed fighter defenses and anti-aircraft artillery to protect their air power resources and industry.[22] The Spanish Civil War underlined the fact that fighter aircraft would play a crucial role in gaining air superiority. The lessons were strong enough to cause Ernst Udet, in charge of production by the late 1930s, to change the projected long run goal for the Luftwaffe's force structure from a ratio between fighters and bombers of 1-to-3 to a ratio of 1-to-2.23

In retrospect, Germany's continental position exercised the greatest influence over her air strategy. Unlike British and American airmen, the Germans *had* to think in terms of land conflict. From the onset of any conflict, the Reich faced a major struggle on the ground, a reality that the Luftwaffe's leaders could not ignore. Thus, air superiority was more than a means to defeat enemy air forces or to attack his factories and cities. It would also enable the Luftwaffe to help the army with close air support[24] and to interdict enemy supply lines to the front.[25]

Germany's geographic position also explains another substantive difference between the Luftwaffe and the U.S. Army Air Forces and the RAF. At the war's outbreak, the Germans believed that the Luftwaffe's structure could dominate the skies over their frontiers. In a sense, they were correct. The Luftwaffe did possess sufficient aircraft to achieve air superiority within the limited geographic framework of Central Europe. In another sense, that geographic frame of reference placed severe intellectual limitations on the ability of the Luftwaffe's leaders to conceptualize the problems associated with war on a continental scale. American and British airmen, the former familiar with the continental distance of the United States, the latter with those of the Empire, thought within a wider framework. Consequently, when the Germans pushed beyond their frontiers, north to Norway, west to France and the Atlantic, South to North Africa and the Mediterranean, and east into the depths of Russia, they discovered themselves out of their depth. What had been sufficient quantitatively and qualitatively to gain air superiority around their frontiers proved insufficient to handle the problems associated with a continental air war.

In the vast spaces of the whole European continent, the Luftwaffe of 1941 and 1942 possessed virtually the same force structure that it had possessed in 1939 and 1940 but now found it virtually impossible to establish anything more than a local air superiority. By the time the Germans had realized the full dimensions of their error it was too late. Admittedly, the Luftwaffe was working in the late thirties to produce a "continental" bomber (one with the load and range of British and American four-engine aircraft).[26] Moreover, the Luftwaffe also developed a long range fighter, the Bf 110, to support its bomber formations deep in enemy territory. That two-engine aircraft, however, proved inadequate for the mission (even to the extent of being unable to protect itself) against first class enemy fighters. On the other hand, the Bf 109, one of the two best air superiority fighters in the world at the end of the 1930s (the other being the British Spitfire), was a very short-range aircraft. The Germans also fell into the same trap as British and American airmen in believing that no single-engine air superiority fighter could achieve sufficient range to accompany deep penetrating bomber formations.[27] Surprisingly, the Luftwaffe did experiment successfully with drop tanks to extend fighter combat radius during the air operations in Spain. That success apparently had no impact on the engineering and operational establishment back home, and in the early years of the war the Bf 109 fought with severely restricted range.[28] This error resulted from an inherent belief that bomber formations could defend themselves (common to all air forces before the war) as well as from an unjustified confidence in the anticipated capabilities of the Bf 110.

German participation on the Nationalist side of the Spanish Civil War provided the Luftwaffe with valuable experience and lessons on future aerial combat. Nevertheless, the size of the German commitment and the nature of the war itself largely confined that experience to the tactical and technical spheres.[29] To begin with, Hitler was unwilling to commit more than a small force to Franco's aid. As he explained in late 1936, it was to Germany's advantage that the Nationalists not win quickly and that Spain divert Europe's attention from the Reich's growing power.[30] Consequently, German air support to Franco remained at a relatively low level throughout the struggle (6,000 men in 1937). At its peak in 1937 the Condor Legion contained no more than forty He 111s, five Do 17s, three Ju 87s, forty-five Bf 109s, four He 45s, and eight He 54s.[31] But the Germans were able to learn a good deal from the experiences of that small force. First of all, in 1936 they recognized from combat how technologically deficient were the first generation of German aircraft sent to Spain. Not only did they rapidly replace those aircraft with newer models such as the Bf 109, but the experiences in Spain sped the process of reequipping the Luftwaffe back home with a new generation of fighters and bombers.

In respect to air superiority, the lessons learned from air- to-air combat proved equally valuable. Like the RAF, the Luftwaffe before the Spanish Civil War had evolved a set of fighter tactics based on close formations of three aircraft. Combat experiences in Spain, however, underlined the vulnerability of such tactics. The future World War II ace, Werner Mölders, established a looser combat formation based on the finger formation of four aircraft with two sections of two aircraft. The German tactics were later copied by nearly all the world's air forces. Mölders, after his return from Spain, wrote a lengthy report on his experiences and that report formed the basis for German air-to-air doctrine at the outbreak of the war.[32] It was to give the Germans an important initial advantage over their opponents.

The finger four tactics proved to be the basic building block of World War II air-to-air combat. They provided not only better visual coverage, but provided direct defensive coverage for those fighters carrying out attacks on enemy aircraft. Equally important was the fact that the Luftwaffe was able to shuttle a considerable number of its senior and particularly its middle level commanders through Spain so that combat experience gained in the Civil War could be passed as widely as possible through the rapidly expanding Luftwaffe.[33] Nevertheless, one should not overestimate the Spanish learning experience and its impact on the German military. On the ground and in the air the commitment of German forces remained limited; military operations in terms of the equipment and tactics were at best primitive; and the air war remained almost entirely out of contact with the rapid development of technology (such as radar) in the advanced industrial nations.

German plans and preparations in 1938 for *Fall Grün*, the invasion of Czechoslovakia, established a pattern that the Luftwaffe repeated in its next two military confrontations: *Fall Weiss* (invasion of Poland in 1939) and *Fall Gelb* (the offensive of May 1940 into France and the Low Countries). The Wehrmacht was so weak in 1938 that German planners had to concentrate nearly all their forces, including the Luftwaffe, on the destruction of the Czech Republic.[34] Virtually nothing remained in the west (five divisions and a smattering of air units) to protect against a possible French reaction.[35] Two air forces, the First and the Third, received the task of destroying the Czech air force and supporting the invasion. The Luftwaffe's plans called for strikes against Czech air fields to destroy the opposing enemy air force and its infrastructure at the outset.[36] The first goal was to establish air superiority over Bohemia and Moravia; destruction of Czech bases and airfields would also prevent

Soviet air reinforcements from reaching the Czechs. After achieving air superiority, the Luftwaffe would then have supported the army's effort with interdiction and close air support strikes as well as attacks on armament industries.

German documents suggest that even the relatively limited military assets of Czechoslovakia presented a considerable problem to the Luftwaffe. First, the German air force was significantly weaker in 1938 than later in September 1939 or May 1940. The numerical change between September 1938 and May 1940 was substantial; the qualitative improvement was also notable. In 1938 the Luftwaffe was still introducing a new generation of aircraft and as late as August was having severe difficulties in maintaining an "operationally ready rate" over 50 percent.[37] Moreover, aircrew transition from obsolete biplane fighters such as the Arado Ar 68 into the Bf 109 had proved to be hazardous, for the high performance and narrow undercarriage of the 109 were difficult to handle. The result was a high accident rate in the fighter force throughout 1937 and 1938.

Luftwaffe staff officers felt that Germany had been lucky in avoiding a war over Czechoslovakia during 1938. A study in fall 1938 pointed out that

In the last months the following special measures have had to be carried through at the same time: 1) [the provision of organizational] equipment to many new units, 2) the reequipment of numerous units [with new aircraft], 3) the early overhaul of about 60 percent of the frontline aircraft, 4) the replacement of spare parts [in squadrons reequipping with new aircraft], 5) rebuilding of numerous aircraft in the supply depots, 6) rearmament of many aircraft [with new weapons], 7) accelerated introduction of overhauled motor models . . . 8) establishment of four new air groups and one new airfield . . . 9) preparation and resupply of mobilization supplies, corresponding to the newly established units, rearmed units, and transferred units . . . The compression of these tasks into a very short time span has once more and in clear fashion pointed out the known lack of readiness in maintenance of flying equipment as well as in technical personnel.[38]

If there were problems in maintaining the Luftwaffe in 1938, prospects on the operational side were equally gloomy. German plans detailed most of the Luftwaffe for operations against Czechoslovakia and left little to defend western airspace or to cover the minuscule deployment of ground forces on the *Westwall*. Even the Czech air defense system posed a substantial challenge. Third Air Force claimed that its air campaign could have crippled Czech air power, but admitted that a combination of inexperienced air crews and bad weather would almost certainly have caused debilitating losses through ground accidents, crashes, and midair collisions.[39] First Air Force, deployed in Saxony and Silesia, reported that while the Luftwaffe enjoyed considerable superiority in aircraft, its airfields had been vulnerable to counter-air operations.[40] Moreover, the Germans feared that the strong anti-aircraft defenses around the fortified zones and industrial centers in Bohemia and Moravia could have inflicted serious losses on German aircraft over Czech targets, especially as low level strikes (vulnerable to anti-aircraft fire) would have formed the basis for most mission profiles during bad weather.[41]

For the Luftwaffe, Hitler's decision not to push the Czech crisis over the brink came as an enormous relief. It avoided war with Britain and France—a war that the Reich would have fought at considerable disadvantage.[42] Unfortunately for the Allies, the Germans used the eleven months between Munich and the outbreak of war far better than their future opponents. By late summer 1939 the Luftwaffe was in considerably better shape than it had been in 1938. This time Hitler refused to be cheated of an opportunity to wage a limited conflict.

Luftwaffe planning for the Polish campaign began in April 1939. Reacting to a British guarantee to Poland, Hitler announced to his entourage that he would "cook the British a stew on which they would choke."[43] He also demanded that the military begin preparations for an attack on Poland at the end of August 1939. Poland presented an easier problem to German planners than had Czechoslovakia in 1938. Not only was the Wehrmacht in better shape,[44] but Poland's strategic situation was even more hopeless than Czechoslovakia's. Hostile territory surrounded the Poles; they possessed no natural defenses; their military forces were less well equipped than the Czechs; and Polish terrain proved an ideal testing ground for the army's mechanized and motorized formations.

Within the strategic context of the decision to conquer Poland, the

Luftwaffe cast its plans. The initial target of air operations would be the Polish air force in a move to gain general air superiority. That would enable the Germans to attack the mobilization and deployment of the Polish army as well as its logistical system.[45] The Germans also planned a massive aerial assault on Warsaw to destroy military and industrial targets in order to paralyze the Polish government at the beginning of hostilities. Bad weather around Warsaw in the early morning of 1 September, however, prevented the Germans from launching such a blow and limited the initial efforts to attacks on the Polish air and ground forces. Once the weather had cleared, air operations against the Polish military were going so well that air force commanders hesitated to shift the emphasis of their attacks to strategic targets.[46]

Despite obsolete aircraft, the Poles proved themselves surprisingly tenacious opponents in the air, and the high skill level among their pilots made them dangerous adversaries.[47] The Poles had deployed a substantial portion of their fighters and bombers to satellite fields before the war broke out, so initial German strikes did not substantially affect the Polish force structure.[48] However, combat against Luftwaffe formations that possessed qualitative and numerical superiority in aircraft soon shredded the Polish air force. Given the German superiority the results were never in question. By the end of the first day the Germans had gained general air superiority; by the end of the second day little remained of the Polish air force support structure.[49] Having gained air superiority, the Luftwaffe finished off the Polish fighter force and shifted its attention to deep interdiction attacks on the enemy's transportation system and to direct support of the ground offensive.

Air superiority allowed the Luftwaffe to accomplish a number of important missions. On the ground the combination of rapidly advancing mechanized units and air strikes against the Polish army proved devastatingly effective. Interdiction missions made it impossible for the Poles to patch together a new line of resistance once German armored forces had broken out into the open. Along the Bzura River, the Luftwaffe caught large Polish formations attempting to fight their way through German encirclements in order to reform along the Vistula.[50] German losses against the Poles were not light. By the time the campaign was over the Luftwaffe had lost forty-seven Bf 109s (5.6 percent of the force structure), eighty-one bombers (6.5 percent), fifty close air support aircraft (13.2 percent) and a total of 261 of all types (7.2 percent). Losses on the western front to the French and British in September speak volumes for the complete lack of activity by Allied air forces. The Germans lost thirteen aircraft in combat and eighteen aircraft through noncombat causes in the west for the entire month of September.[51]

The German ability to maintain the thrust of mechanized units and to push fighter and close air support coverage forward rested on an effective air transport system. Airlift squadrons, equipped with Ju 52s, resupplied the army as mechanized forces outstripped the ground based logistic system. At the same time Bf 109 squadrons established themselves on forward airfields within conquered territory and received supplies of fuel, ammunition, and parts through airlift. This operational concept, established before the war, played a crucial role in helping the Luftwaffe's fighter force keep up with the army's thrusts during both the French campaign and the invasion of Russia as well as in Poland; such a system, however, had no utility in circumstances where no forward movement occurred and where opposing air forces grappled independently of ground operations.[52]

The German victory over Poland represented only a limited operational success, however. The entrance of Britain and France into the war placed Germany in a dangerous strategic situation. Hitler had, in fact, underestimated the Reich's economic and strategic vulnerability, while hoping that the Soviet Union could make up whatever shortages an Allied blockade caused.[53] As a result of serious economic difficulties, Hitler pushed for an immediate offensive against the West; the army, unhappy with the performance of its troops in Poland, argued strongly against offensive operations before spring. Generally, the Luftwaffe seconded these army efforts, but for different reasons. First, it preferred to wait out the bad weather. Second, a rest period after Poland allowed it to make good its losses, as well as to build up fighter and bomber strength, quantitatively and qualitatively (in the latter case with the introduction of the Ju 88 into the bomber force). In any case, bad weather and the unwillingness of Anglo-French strategists and politicians to put *any* pressure on the Reich allowed the Germans to postpone the western offensive until spring 1940.[54]

During the winter Hitler and the German navy began to worry about the economic and strategic vulnerability of Scandinavia, particularly the ore traffic along the Norwegian coast during the winter and spring. The *Altmark* affair in February 1940 (in which British destroyers attacked a German supply ship and freed Allied merchant sailors on board) convinced the Führer that if he did not act soon, the British would block the transshipment of iron ore. Therefore, he ordered the Wehrmacht to plan an invasion of Denmark and Norway ("*Weserübung*"). For one of the few times in the war there was a modicum of interservice cooperation in Germany under the OKW.

For the Luftwaffe and the other services, Denmark presented little problem, but Norway was another matter. Her long coast and ready accessibility to British sea power made military operations against her hazardous. In retrospect the Luftwaffe played *the* critical role in operations against Norway. At the start the Germans faced a vacuum in terms of Norwegian air defenses. Facing no opposing air force, the Luftwaffe's success in establishing air superiority depended upon whether the Germans could gain and hold the Norwegian airfields. If they could not do so and if the Norwegians held on, then the RAF could move into Norway and with the Royal Navy isolate German forces that had landed in the ports.

The seaborne landings went according to plan with one important exception. German naval forces, moving up the Oslo fjord, ran into significant resistance and lost the heavy cruiser *Blücher* and with it not only the landing force but also much of the administrative structure assigned to the occupation. Had the Norwegians acted with dispatch, they could have mobilized, protected Oslo's main airport with reservists, and denied the Germans access into the heart of their country. They did not. Informed by the German operations officer in charge of the invasion (who was then in Oslo to threaten the Norwegian government) of the difficulties encountered by the navy during the move up the Oslo fjord, the Luftwaffe had aircraft over Oslo harbor within an hour. By 0900 hours a small element of German paratroopers had seized the airport. Troops, rushed in by Luftwaffe Ju 52s, allowed the Germans to seize the capital by early afternoon.[55] To the west of Oslo, airborne troops seized the Stavanger/Sola airfield and by the end of the day 180 German aircraft had landed (including bombers, twin-engine fighters, and Stukas). Luftwaffe transport squadrons had flown in not only fuel, ammunition, and maintenance personnel but light Flak units as well. By the next day, other Norwegian airfields from Trondheim to Stavanger had fallen into German hands.[56]

The rapid establishment of German air power in the vacuum of central and southern Norway won the campaign. From its new bases, the Luftwaffe dominated the land and naval battles. Not only did it protect German forces from the RAF, but it prevented a timely and effective intervention by Allied sea power.[57] The speed with which the Germans had seized the airfields and then turned them into operational bases capable of supporting significant air operations was one of the nastiest surprises of the campaign. Once the Luftwaffe had the fields in operating condition, it was able to isolate the battlefield, to support the ground forces in breaking up what was left of the Norwegian army or what the Allies managed to land, and to supply German units throughout the theater.[58] The results were then never in doubt.

After the opening move in Scandinavia, the Wehrmacht launched its massive spring assault on Western Europe. Two great German air forces (the Second and the Third) covered the movement of three army groups, seven armies, and 136 divisions (ten of them panzer).[59] Altogether the Luftwaffe deployed over 3,500 aircraft out of a frontline strength of 4,500.[60] (The remainder supported Luftwaffe operations in Norway and the training establishment.) In addition, a transport command of 475 Ju 52s (refitted after the Norwegian campaign) provided airlift for extensive airborne operations against the Dutch and Belgians. In the long run the Ju 52s formed the logistical backbone for the rapid forward deployment of air units, particularly the short range Bf 109s, as the battle surged deeper into Allied territory.

At the start of the campaign the Luftwaffe aimed to achieve air superiority over the battlefield by attacking Allied air bases and aircraft; to provide airborne drops on Dutch and Belgian forts and bridges; and to support the army along the main axis of its advance through the Ardennes, across the Meuse and on toward the English Channel. The Luftwaffe executed the second of these objectives with dispatch, although with heavy casualties. The paratrooper and glider assaults on Fort Eban Emael and Dutch bridges, along with the hammering advance of the infantry in Army Group B, fixed Allied attention on the north and the seeming replay of the 1914 Schlieffen Plan. And all the while, German armored and motorized forces rolled through the dark forests of the Ardennes on their way to Sedan and other points along the Meuse.

Meanwhile, the Luftwaffe launched a series of blows at Allied air bases to disrupt and destroy the infrastructure of Allied air power. Unsupported bomber formations mounted those attacks and, as was the case throughout the war, they paid a heavy price.[61] Luckily for the Germans, the French air force, although performing substantially better than most historians have acknowledged, faced insurmountable problems. As with the Luftwaffe in 1937 and 1938, and the RAF in 1938 and 1939, the French in 1940 were making the transition to a new generation of aircraft. As a result, many French fighter squadrons in early 1940 were running operational ready rates of barely 40 percent. The pressures of combat operations only exacerbated these difficulties.[62] Thus, while new fighter aircraft possessed considerable potential, they did not provide the operational performance (in terms of sorties) of the Spitfire, Hurricane, or Bf 109.

The massive German bomber and fighter strikes at Allied air bases in the Low Countries and France rocked Allied air forces back on their heels and placed them in a defensive posture from which they never recovered on the continent. Because the Germans enjoyed the initiative, they were able to gain a considerable measure of surprise. It was not that those in the west did not expect a German attack. It was rather the intensity and ferocity of the assault on military organizations that, whatever their expectations, were existing in a peacetime environment.

German airborne drops on the Dutch airfields, while they did not succeed in gaining immediate operational control of the bases, rendered the Dutch air force *hors de combat* at the outset. Luftwaffe attacks on Belgian airfields destroyed approximately half of the Belgian aircraft on the ground and damaged the support structure extensively. The Germans also managed to inflict substantial damage to some British and French airfields. At Cond Vraux, the base of the RAF's 114 Squadron, Luftwaffe bombers destroyed six of eighteen aircraft and damaged the remainder severely enough to render them unserviceable. Attacks on other British airfields were not as successful.[63] Nevertheless, the outset of hostilities had been anything but favorable to Allied air operations.

Substantially adding to Anglo-French difficulties was the fact that their command and control system failed to function on 10 May. Not until 1100 on that day did Allied air commanders receive authority to attack German columns and airfields and then with the admonition that they were "at all costs [to] avoid bombing built up areas."[64] The delay in releasing the air forces to attack even military targets reflected the failure of Allied political leadership to act decisively even when confronted with the terrible reality of the German opening moves. By and large, the RAF's aircraft in France confined themselves to flying cover for the BEF as it moved into Belgium. They saw little of the Luftwaffe, largely because the Germans had no intention of interfering with a move that so obviously played into their hands.

Unfortunately, almost all of the Luftwaffe's operational records were destroyed at the end of the war, and it is impossible to determine exactly how the Germans allocated their air resources for the campaign in the west. Nevertheless, the overall conduct of the first weeks of the campaign do suggest a general pattern to German air operations. The first strategic objective of Luftwaffe operations was to destroy or at least severely impair the Allied air base structure, thus rendering it difficult for the enemy air forces to intervene against the forward movement of the German army. These German bomber and long-range fighter attacks do *not* seem to have received substantial support from the Bf 109 force. Rather the German single-engine fighter force seems to have largely been engaged in screening and protecting the movement forward of the armored force through the Ardennes. Some of the fighter force also engaged in straight out air-to-air missions to seek out and destroy enemy fighters and bombers.[65] Neither the fighters nor the twin-engine bomber force engaged in close air support missions for the army. Only the Stuka force flew that mission profile and then largely in support of the breakthrough along the Meuse on 13 May. After the breakthrough at Sedan, the Stuka force reverted to the air interdiction mission, as it possessed very limited capability in 1940 to perform close air support in a mobile environment.[66]

At the same time that the Luftwaffe's bombers and fighters were striking Allied airfields and the support structure, Bf 109s were making an intense effort to sweep the skies over the Ardennes of Allied reconnaissance aircraft and bombers that might give away the main direction of the *Schwehrpunkt*. Only the RAF appears to have made a sustained effort in the Ardennes region, and strong Luftwaffe forces in the area inflicted crippling casualties on the British. On 10 May, four waves of Battle bombers covered by six Hurricanes attempted to strike German columns in the Ardennes. Of thirty-two Battles, the Germans managed to shoot down thirteen and damage the remaining nineteen. On the 11 May, eight Battles again attacked the Germans in the Ardennes. Of the attacking aircraft, only one returned badly damaged; the remainder had all been lost.[67]

Its massive air operations cost the Luftwaffe heavily. On 10 May the Germans lost eighty-three aircraft (including forty- seven bombers and twenty-five fighters)—more aircraft than they would lose on any day of the Battle of Britain. On the following day, the Germans lost a further forty-two aircraft (including twenty-two bombers, eight dive bombers, and ten fighters.)[68] Allied losses were no less severe,[69] but, of course, the Germans enjoyed a considerable quantitative advantage over both opposing air forces (which was magnified by the fact that a substantial portion of the RAF had remained in the British Isles for air defense purposes and was consequently not involved in the battle for air superiority over the western front).

On 12 May Guderian's panzer divisions began crossing the Semois. Allied air attacks, especially by Battle bombers, caused the Germans considerable difficulties, including forcing Guderian to move his head-quarters. Defending German fighters and anti-aircraft guns inflicted heavy casualties on RAF bombers, shooting down eighteen out of fifty aircraft. The French were able to inflict some painful casualties on the Luftwaffe. Five Curtis fighters caught twelve Stukas returning unescorted from a raid and shot all of them down.[70] Unfortunately, the general air superiority that the Germans enjoyed over the Ardennes made such incidents very much the exception.

By 13 May German armored forces had come up on the Meuse between Dinant and Sedan. By the 14 May Guderian had his infantry, supported by artillery and Stukas, across the river and busily engaged in punching through French defenses. Even more important was the fact that the Germans had managed to bridge the Meuse and begin moving armored forces across the river. The threat posed by this German thrust and the collapse of French units produced by the Luftwaffe's Stukas and German infantry finally awoke the French high command to the danger. Desperate calls from the French led the RAF to throw its bomber forces against the growing German penetration. The effort aimed at destroying the bridges thrown across the Meuse by German combat engineers and at attacking German columns moving up and across the river. The results were a disaster for the RAF. Luftwaffe fighters and anti-aircraft savaged attacking formations. The official history records the RAF's losses as thirty-five out of sixty-three Battle bombers dispatched and five out of eight Blenheim bombers dispatched (or forty out of seventy-one, 56 percent of the attacking force).[71]

In these operations the RAF caused the Germans serious difficulties. The war diary of XIX Panzer Corps (Guderian's force) noted that "the completion of the military bridge at Donchery had not yet been carried out owing to heavy flanking artillery fire and long bombing attacks on the bridging point . . . . Throughout the day all three divisions have had to endure constant air attack—especially at the crossing and bridging points. Our fighter cover is inadequate. Requests [for increased fighter protection] are still unsuccessful."[72] Luftwaffe reports also indicated the pressure that Allied air attacks were exerting in the Ardennes: "vigorous enemy fighter activity through which our close reconnaissance in particular is severely impeded."[73] Nevertheless, while Guderian's war diary exhibited dismay over inadequacies in the fighter cover, German air defenses had destroyed 56 percent of the attacking RAF bombers. No air force could support such a level of attrition, and on the next day the Germans noted a significant decrease in the intensity of air attacks along the perimeter of the breakthrough.

Once German armored formations had broken through French defenses along the Meuse, the campaign was over. The French army, frozen in a doctrinal rigidity of its own making, was incapable of replying to the German thrust. Exploitation of the breakthrough now proceeded with dispatch. Behind surging columns of armored and motorized units, the Luftwaffe pushed its operating bases forward so that Bf 109 and dive bomber units could remain in contact with the ground forces that were rapidly pushing ahead and in danger of passing out of the range of effective air cover. On 17 May, within twenty-four hours of its abandonment by the French, German fighter squadrons had established themselves at Charleville, west of the Meuse. Because the army's logistical system was choking the Meuse bridges, Ju 52 transports flew in everything from maintenance personnel to fuel and munitions. So short of POL was the forward operating base that aircraft returning from Charleville to rear area bases had all but the minimum fuel load pumped out of their tanks.[74]

Once in the open, the Germans found a noticeable slackening in the Allied air resistance. Anglo-French air units scrambled pell mell to the south as the German army chewed through their frontline bases. Ground crews, supplies, and maintenance equipment all had to move south of the Somme with little warning, and the process of sorting out ground organizations in the wreckage of defeat represented an impossible task, given the available time. In addition to problems posed by the rapid move to the south, German air attacks placed considerable pressure on the Allied support structure as well as on fighting strength. The one group of Devoitine 520s (the newest and best French fighter aircraft—close in performance to the Bf 109s and Spitfires—put up a respectable showing in air-to-air combat with the Luftwaffe, but by 21 May had lost half of its aircraft on the ground through German attacks on its bases.[75]

The Luftwaffe met its first setback over Dunkirk. There the Germans faced an enemy who possessed first class equipment and whose base structure remained intact and invulnerable to ground operations. Fighter Command had not committed any of its limited number of Spitfires to the defense of France. Thus, while the Hurricanes and Devoitine 520s had put up a respectable showing in air-to-air combat, only over Dunkirk did the Luftwaffe run into aircraft fully the equal of the Bf 109. Moreover, British bases on the other side of the Channel lay closer to evacuation beaches than did such German forward operating bases as Charleville. Consequently, British fighters possessed more loiter time in the combat zone than did the Bf 109s. Given German numbers, the result was a furious air battle in which RAF Fighter Command thwarted Göring's promise that the Luftwaffe, by itself, could destroy the trapped Allied ground forces.[76] With the 109s at the outer limit of their range and with the bomber force still flying from bases in Western Germany, the coordination of the two was a formidable task.

By 26 May Fighter Command was providing almost continuous cover over Dunkirk. Standing patrols of squadron strength (ten plus fighters) were taking off from British airfields approximately every fifty minutes. Because the Bf 109s were operating at the outer limits of their range, the RAF was able to interfere significantly with German operations. While some German bomber formations received fighter escort, others did not. On 26 May the British lost only six fighters while the Luftwaffe lost thirty-seven, the great majority in the Dunkirk area.[77] May 27 told a similar story. Sixteen squadrons of Fighter Command covered the Dunkirk area with pilots flying two to three missions each.[78] *Fliegerkorps* II, engulfed in the fighting over Dunkirk, reported that it had lost more aircraft on the 27th than in the previous ten days of the campaign.[79] The battle for air superiority in the skies over Dunkirk was costly to both sides. From 26 May through 3 June, the RAF lost 177 aircraft destroyed or damaged; the Germans lost 240.[80] The air battle was in and by itself inconclusive. Neither side had won a clearcut victory. Nevertheless, strategically Fighter Command was able to oppose the Luftwaffe successfully and thwart the Germans from bringing the full weight of their air power to bear on the evacuation. In that sense, "the miracle of Dunkirk" was as much Fighter Command's victory as it was the victory of the Royal Navy and the little boats.

Dunkirk, as the fighter ace Adolf Galland suggested in his memoirs, should have alerted the German high command to the inherent weaknesses in the Luftwaffe's force structure.[81] The Germans possessed the range and striking power to gain air superiority, provided air operations occurred within a limited space and where the army forward thrusts could extend aircraft range by seizing bases for further operations. Whether the Luftwaffe could defeat an air force whose bases were not threatened by ground operations and who possessed a level of production equal to if not superior in some categories was another matter.

The current conception of the defeat of France is that it cost the Germans relatively little. The German victory in France often serves as the paradigm for the mobile, flexible operations advanced by many of the current critics of the American defense establishment. The cost, therefore, is worth noting. In a campaign during which heavy fighting occurred over less than a period of a month, German and Allied casualties added up to over half a million (not including prisoners of war). German panzer forces, moreover, lost nearly 30 percent of their tanks (753 out of 2,574) during the furious advance.[82] The Luftwaffe suffered equally. In May alone it lost 27.4 percent of its bomber force, 12.3 percent of its fighter force and 20.2 percent of its total force structure.[83] Table 7.1 gives the extent of German losses from the whole of the May-June 1940 period.[84] Those losses suggest, as do ground casualties, that the French put up a more respectable showing than historians have allowed. The losses also suggest that the defeat of 1940 was due less to national rot than to gross incompetence of France's military leadership.[85]

The RAF's fighter losses during the French campaign amounted to 474 aircraft—more than half the number of fighters with which it had begun operations on 10 May (in England as well as in France).[86] German pilot losses among Bf 109 forces were not disastrous but do suggest the intensity of the fighting. Records indicate that fighter pilot losses for the campaign were 15.2 percent of the pilots on active service at the onset of operations in the west.[87]

TABLE 7.1

German Aircraft Losses (Destroyed), May–June 1940
Destroyed on Operations

Type of Aircraft	Strength 4.5.40.	Due to Enemy Action	Not Due to Enemy Action	Total	Destroyed Not on Operations	Total Destroyed	Losses as Percent of Initial Strength
Close recce	345	67	5	72	6	78	23
Long-range recce	321	68	18	86	2	88	27
Single-engine fighters	1,329	169	66	235	22	257	19
Twin-engine fighters	367	90	16	106	4	110	30
Bombers	1,758	438	53	491	30	521	30
Dive bombers	417	89	24	113	9	122	30
Transport	531	188	18	206	7	213	40
Coastal	421	20	16	36	3	39	16
Total	**5,349**	**1,129**	**216**	**1,345**	**83**	**1,428**	**28**

The campaign in France brings out several interesting points. The German effort in the air and on the ground generally saw close integration and cooperation at all levels. The sum of that cooperative effort resulted in a devastating military victory over Allied military power in the west. The Luftwaffe did not gain complete air superiority over its opponents at the outset of the campaign. However, from 10 May on, the pressure that it placed on opposing air forces on the ground and in the air allowed it to carry out its mission objectives, while generally preventing its opponents from executing theirs. The value that general air superiority contributed to the German victory is best represented by events along the Meuse between 13 and 15 May. From the outset the Luftwaffe was able to shield close air support attacks on French troops on the left bank of the Meuse from Allied interference. The incident,

mentioned above, when five Curtis fighters of the French air force intercepted twelve Stukas and shot down all of them suggests what stronger
Allied fighter forces might have been able to do. The result of such
Allied weakness was that the German Stuka forces laid down a devastating pattern of support, materially contributing to the collapse of French
troops along the Meuse, especially at Sedan in front of Guderian's XIX
Panzer Corps. Then, when the Germans had broken through, Allied air
power attempted to cut the Meuse bridges with a sustained bombing
effort. Allied fighters, inferior in numbers, were never able to give their
bombers adequate support. The result was an aerial massacre. While
Allied bombing attacks did cause the Germans severe discomfiture, they
were not sustainable. As a result, there was hardly anything left after
15 May to carry out further heavy attacks.

Once into the open, German armored forces cut a wide swath
through the rear area of northern France, and forced Allied air forces to
abandon their bases. That hurried retreat resulted in the loss of spares,
parts, ammunition, and fuel, all of which was in short supply on new
and unprepared airstrips. The German ground advance also thoroughly
disrupted the command and control system (which had never worked
particularly well from the point of view of air commanders). From that
point on, the Allied air effort against the Germans in France rapidly
diminished.

The air battle over Dunkirk was a different story. There the German
army had outrun its air support and Luftwaffe efforts to dominate the
battle area faced insurmountable obstacles. Even with the forward
movement of Luftwaffe fighters behind surging army spearheads, Bf
109s, the heart of the air superiority force, remained far from the evacuation beaches. On the other side of the Channel, Fighter Command,
flying from secure bases, not under the threat of ground operations,
was able to disrupt the Luftwaffe's effort to halt the evacuation. The
escape of the British army, in effect, made viable the strategic defense
of Great Britain that summer. In that sense the RAF won an important
victory by preventing unhindered use of the Luftwaffe's capability.

In the final analysis Allied air forces were insufficient to thwart the
combined German effort on the ground and in the air. Nevertheless,
the historian leaves the French campaign with the sense of how close
the German advance had come to serious difficulties. One can argue
that the decision that lost the air battle and perhaps the campaign was
taken in October and November 1938. Under great pressure both in the

House of Commons and from the public to repair the glaring deficiencies in British defenses, the government of Neville Chamberlain announced major increases in its purchasing plans for fighter aircraft. However, it was all a sham by a government determined not to spend any more on national defense. What Chamberlain and his advisers did was to extend the contracts for Hurricanes and Spitfires without increasing monthly production totals, and there was no net gain over what was already planned. The production performance of the British aircraft industry from late 1938 through summer 1940 clearly indicates that monthly production figures could have been substantially increased.[88] They were not and thus Allied air forces were quantitatively and qualitatively inferior in May 1940. What Allied air forces with 300 to 400 more Spitfires and Hurricanes might have achieved is obvious. Not only would they have been able to contest air superiority with the Luftwaffe for a longer period of time, but they could have protected their own bases better and provided significantly more support for bomber sorties. Furthermore, the long lines of vehicles curling up towards the Meuse or the traffic jams of vehicles waiting to cross provided wonderful targets. The 56 percent loss of British bombers on 15 May, however, rendered the strike force ineffective after one mission and the movement forward of the panzers continued without interference. The result was a disastrous defeat for British and French military forces.

As surprised as others by the completeness of the French collapse, the Germans believed they had won the war. With the armistice, Germany's leaders went on vacation. Hitler spent time in visiting Paris and World War I battlefields as well as picnics along the Rhine.[89] His military advisers did not work much harder. General Alfred Jodl, number two on the OKW staff, suggested at the end of June that "the final victory of Germany over England is only a question of time."[90] Hitler himself hoped right to the end of July that Britain, recognizing her hopeless position, would sue for peace. As early as 20 May he had suggested that England could have peace for the asking.[91]

Considering the abject performance of British policymaking in the late 1930s,[92] one can excuse the Germans for their belief that Britain would surrender. What they missed was that the British mood had substantially changed. Churchill's oratory was not mere rhetoric; it indicated a first-class strategic mind with the ruthlessness and toughness of spirit needed to back it up. On 5 July 1940 after fruitless negotiations, the Royal Navy destroyed much of the French fleet in its North African base at Mers-el-Kebir.[93] That action signaled that Britain was in the war for the duration. Yet, two weeks later, Hitler was still extending the olive branch to Great Britain as he promoted his admirals and generals with great ceremony in Berlin.

The strategic problem posed by British resistance represented a wholly new dimension of strategy to the Wehrmacht's *Weltanschauung*. The Reich's military forces were not only ill-equipped to solve that strategic problem, but they possessed none of the intellectual and professional background that an amphibious assault on the English coast demanded. In fact, one can wonder how seriously the Germans took the proposed invasion, code-named Operation "Sea Lion." The army willfully disregarded the navy's logistical capabilities in presenting plans for a seventy-mile invasion front. Operations off the North Cape in early June best represented the navy's attitude. Despite earlier discussions between Hitler and his naval commander in chief, Grand Admiral Erich Raeder, over a possible landing on the British Isles, the naval high command committed Germany's only two battle cruisers, the *Gneisenau* and the *Scharnhorst*, to operations off the Norwegian coast, more to influence postwar budget debates than for strategic reasons. As a result both were so damaged that neither was ready for operations until December. Thus, Germany had at her disposal only one heavy cruiser, two light cruisers, and four destroyers at the end of June.[94]

Because the army and navy had neither the inclination nor the resources for combined operations, Germany possessed no suitable landing craft in 1940. Consequently, "Sea Lion" rested its cross channel logistical lift on Rhine river barges. With few escort vessels, the Germans had to count on the Luftwaffe to deny the Royal Navy and the RAF access to the channel. Summing up the general sloppiness of German strategic thinking in the summer is an OKW directive, signed by Field Marshall Wilhelm Keitel on 12 July. That directive suggested that the Luftwaffe would have to substitute for the absence of naval power. With air superiority, the landings on the British coast would take the form of a powerful river crossing.[95]

Since the Royal Navy had stationed a large number of destroyers with cruiser support at Portsmouth and Harwich among other locations, one can seriously doubt whether "Sea Lion" ever had a chance even had the Luftwaffe beaten the RAF in September.[96] Even a few British destroyers in among such an invasion fleet would have created a disaster for the Germans. The Luftwaffe could only have intercepted a bare majority of the thirty to forty destroyers and cruisers that the British had already deployed to meet the invasion by mid-August. Moving at speeds upwards of thirty-plus knots, destroyers would have been an extraordinarily difficult target to hit. Moreover, the Luftwaffe and the German navy had done virtually no preparatory work to iron out how they would cooperate in either protecting convoys against enemy surface attacks or how they would cooperate in a massive air-sea battle. It is notable that in 1941 in the waters off Crete, the Luftwaffe could not protect the seaborne landings of German troops from the Royal Navy despite total air superiority and perfect weather conditions.[97] Thus, the Luftwaffe had little sense of the complex tasks that its air units would have faced in subduing the Royal Navy while supporting an invasion.[98]

Interestingly, neither Hitler nor Churchill seems to have fully believed that a cross-channel invasion was in the cards. The prime minister in September 1940 sent a sizeable percentage of Britain's armored strength to the Middle East—hardly the decision of a man who believed an invasion was imminent.[99] Hitler also appears to have been dubious about prospects for the invasion; from the beginning the Führer, for the only time in the war, had little to do with the planning and conduct of operations preparing the way for "Sea Lion."

The Luftwaffe faced very different strategic problems in the summer of 1940 than it had dealt with in its three previous campaigns. Its opponent, the RAF, possessed relatively secure bases that would not be under ground attack unless it was first defeated. Consequently, the Luftwaffe ground support structure could not move forward behind the army's advance. Only air attacks could hope to disrupt RAF maintenance and supply. Nevertheless, with new bases in the Low Countries and northern France, German bombers could reach most of the important transportation, industrial, and population centers in Britain as well as RAF airfields. And unlike other air forces in 1940, the Luftwaffe had attempted to solve the long-range escort problem. Unfortunately for German prospects, the fighter explicitly designed for that role, the Bf 110, while possessing the range to accompany deep penetration missions, could not stand up against first-class, single-engine fighters. Against the Hurricane and Spitfire it lacked both speed and maneuverability—a deadly combination. Thus, the Bf 109 would have to protect not only the bombers but Bf 110 formations as well, and the range of the Bf 109 was such that even with the airfields in Pas de Calais it could barely reach London.

In a June 1940 memorandum, Jodl sketched out the strategic framework for victory over Britain.[100] For a direct strategy, he saw three approaches: an air and naval offensive against British shipping along with attacks against industry; terror attacks against major cities; and finally, landing operations to occupy an already prostrate England. The Luftwaffe, Jodl suggested, must gain air superiority; by destroying industrial plants, it would insure that the RAF could not recover. The strategic study also suggested that air superiority would prevent the RAF from striking the Reich and particularly the Ruhr. It is within this context that German attacks on Bomber Command's airfields must be seen; German air strategy during the Battle of Britain aimed not only at Fighter Command's destruction, but also at the elimination of the bombing threat to Germany.

On the day that Jodl's memorandum surfaced, Göring issued general instructions to his forces.[101] After redeployment to airfields near Britain, the Luftwaffe would go after the RAF. Its targets would be Fighter Command and Bomber Command, ground support echelons, and the aircraft industry. Göring suggested, "as long as the enemy air force is not destroyed, it is the basic principle of the conduct of an air war to attack the enemy air units at every possible opportunity—by day and night, in the air, and on the ground—without regard for other missions." Once the Luftwaffe had succeeded in gaining air superiority, it would assault British imports and stockpiles. The heavy losses of the French campaign had indeed made an impact on the *Reichsmarschall*. He urged his commanders to conserve the Luftwaffe's fighting strength as much as possible, and not to allow overcommitments of either personnel or materiél.

To destroy the RAF and gain air superiority, the Luftwaffe deployed Second and Third Air Forces in France and the Low Countries and Fifth Air Force in Norway. The former two controlled over 2,600 aircraft, while the latter possessed nearly 300 more (see Table 7.2 for the force deployment of the Luftwaffe in mid-July).[102] The redeployment of such large air units from bases in Germany required considerable time and effort. In addition, the Luftwaffe faced difficulties in making good the losses suffered in France. Thus, the two month hiatus between victory over France and the beginning of the air campaign against the British Isles reflected the above factors, as well as German overconfidence.

TABLE 7.2

German Air Strength, July 20, 1940

	Second and Third Air Forces			Fifth Air Force			Luftwaffe		
	Strength	In commission No.	Percent	Strength	In commission No.	Percent	Strength	In commission No.	Percent
Long-range bombers	1,131	769	68	129	95	74	1,401	903	64
Dive bombers	316	248	79	–	–	–	449	332	74
Single engine fighters	809	656	81	84	69	82	1,060	865	82
Twin engine fighters	246	168	68	34	32	94	398	293	74
Long-range recce	67	48	72	48	33	69	280	189	74
Short-range recce	90	–	–	–	–	–	250	178	71
Total	2,659	1,889	74	295	229	78	3,838	2,760	72

German prospects were not helped by their intelligence services. Colonel Joseph "Beppo" Schmid provided the basic survey of the RAF on 16 July.[103] Like much of Germany's intelligence work, Schmid's study was arrogantly overconfident of Luftwaffe capabilities and generally ignorant about the British defense system. Schmid only came close to the mark in the quantitative counters; his estimate calculated that with fifty squadrons the RAF possessed approximately 900 fighters (675 in commission). In fact, the RAF possessed 871 fighters of which 644 were operationally ready.[104] Schmid also got the ratio between Spitfire and Hurricanes generally right, suggesting a forty to sixty ratio. In the operational squadrons the RAF possessed 279 Spitfires and 462 Hurricanes, a thirty-eight to sixty-two ratio.[105]

From there, his estimate went downhill. Schmid characterized both Hurricane and Spitfire as inferior to the Bf 109, while only a "skillfully handled" Spitfire was superior to the Bf 110. He calculated that British fighter production lay somewhere between 180 and 300 machines per month (actual production reached 496 for the month of July, 476 for August, and 467 for September),[106] but argued that production would soon go down due to reorganization, vulnerability to air attack, and raw material shortages.

But Schmid and the Luftwaffe's intelligence service made their greatest errors in evaluating the higher levels of British command and control. A short paragraph dismissed Fighter Command's ability to control its units effectively: "The Command at high level is inflexible in its organization and strategy. As formations are rigidly attached to their home bases, command at medium level suffers mainly from operations being controlled in most cases by officers no longer accustomed to flying (station commanders). Command at low level is generally energetic but lacks tactical skill."[107] Finally, and equally disastrously, Schmid never mentioned the British radar system and its implications for the attacking German forces. The intelligence estimate ended in the confident assertion that

> the Luftwaffe is clearly superior to the RAF as regards strength, equipment, training, command and location of bases. In the event of an intensification of air warfare the Luftwaffe, unlike the RAF, will be in a position in every respect to achieve a decisive effect this year if the time for the start of large scale operations is set early enough to allow advantage to be taken of the months with relatively favorable weather conditions (July to the beginning of October).[108]

Schmid's memorandum is important not only for the gross overconfidence it reflected, but also because such intelligence errors would plague the Germans throughout the war.[109]

A second point needs to be made about the failure to see the implications of the British radar system. While the Germans were somewhat behind the British in technical developments, they did possess radar. However, considering the nature of the war that Germany had prepared to fight (and thus far had fought—a continental air and land war in which their forces were on the offensive), the Germans had not developed an air defense system as sophisticated as that of the British. With little sense of such a system's advantages, the German missed an opportunity to use radar for offensive air operations. Had the *Jafus* (the Second and Third Air Force officers responsible for fighter operations) possessed radar plots of air operations over Britain they might have played a more significant role. Particularly in the early phases of operations, they could have reacted to Fighter Command's responses as attacking forces built up behind Pas de Calais. However, with little active or passive intelligence, the *Jafus* spent a dismal summer in trying to make sense of what returning pilots reported. Even disregarding the advantages that such a system would have given, German radar on Pas de Calais could at least have allowed the Luftwaffe to observe British responses to raids and to understand better the British defensive system.

In fact, the Luftwaffe faced a far more tenacious opponent than it supposed. The British air defense system, led by its commander in chief, Air Marshal Sir Hugh Dowding, represented a well-equipped and well-prepared force. The cutting edge of the defenses was Fighter Command, disposing of roughly 600 serviceable Spitfires and Hurricanes (approximately 800 altogether).[110] While the RAF had suffered heavy losses in May and June (509 Spitfires and Hurricanes), the number of fighters available was never a serious problem.[111] British fighter production had already overtaken Germany's, and with Lord Beaverbrook's demanding pressure the ratio between national figures would be nearly two to one in favor of the British by late summer.[112] The greatest problem then, was not lack of aircraft, but lack of pilots. The RAF had lost nearly 300 pilots in France, and at the beginning of July out of an establishment of 1,450 pilots, Fighter Command possessed only 1,253. Moreover, most pilots lost over France had been experienced; their replacements were straight out of the OTUs (Operational Training Units).[113] In microcosm the strategic problem confronting the Germans in the summer was similar to that facing Allied air forces in 1943, particularly in terms of the daylight offensive. Because of the Bf 109's limited range, German bombers could only strike targets during the day in southern England, where fighter protection could hold losses to acceptable levels. This situation allowed the RAF a substantial portion of the British Isles as a sanctuary. Within that area, relatively free from the threat of air attack, the RAF could establish and control an air reserve and protect a substantial portion of Britain's industrial production, particularly in the Birmingham-Liverpool area. Should the Germans attempt to attack targets behind London, the RAF could impose an unacceptable level of attrition on the unescorted bombers.

Moreover, the limited range of German fighter cover allowed the British one option that they never had to exercise. Should pressure on Fighter Command become too great, it could withdraw north of London to refit and reorganize; then if the Germans launched "Sea Lion" it could then return to the struggle with full force. Consequently, the Luftwaffe could only impose on Fighter Command a rate of attrition that its commanders would accept. The Germans were never in a position to attack the RAF over the full extent of its domains. Similarly in the 1943 daylight air battles, American escort fighters could only protect the B-17s and B-24s on a line running roughly along the western bank of the Rhine. Beyond that protective curtain, the Luftwaffe's fighters imposed an unacceptable attrition rate on the Americans. Not until Eighth Air Force possessed escort fighters of sufficient range to reach over the length and breadth of the Reich were Allied air forces able to break the Luftwaffe and to win a general air superiority over the European continent.

The pause after the fall of France reflected not only considerable German overconfidence and the belief that Britain would recognize her hopeless situation, but also the organizational and logistic difficulties of shifting to new bases along the Channel. Moreover, the Germans had to make good aircraft losses suffered in the spring and integrate new crews into active units. Yet a pervasive mood of overconfidence marked the German approach to the battle. Operational estimates forecast that only four days of major operations over southern England would break Fighter Command. The Luftwaffe would then need only four weeks to eliminate the remainder of RAF and to destroy the factories on which British air strength rested. Then, the Luftwaffe, savaging British cities by day and night, could protect the "Sea Lion" landings, if required to give, as Jodl characterized the operation, a final "death blow" (*Todesstoss*).[114]

By the end of July, Luftwaffe thinking for the coming air battle had crystallized. On 21 July Göring suggested to senior commanders that besides the RAF, the British aircraft industry represented a critical target for winning air superiority. Above all, the *Reichsmarschall* urged, they must aim to weaken Fighter Command's morale and strength. In contrast to his late August directions, the *Reichsmarschall* argued that the fighter forces should possess maximum operational latitude in protecting bomber formations. Thus, Luftwaffe bomber raids would bring up the RAF's fighters and fighter sweeps would seek out and attack the Spitfire and Hurricanes wherever they could be found: on the ground, taking off, climbing to fighting altitude, or attacking German bombers. And the Bf 109s would enjoy the advantage of the initiative, since they were not tied exclusively to protecting the bombers. Such a strategy would maximize fighter speed and maneuverability.[115]

Three days later *Fliegerkorps I* mapped out four basic missions for the Luftwaffe in the upcoming campaign. The foremost task was to gain air superiority through attacks on the RAF and its supporting aircraft factories, particularly those producing engines. Second, the Luftwaffe would support the future invasion with attacks on the enemy bomber force and fleet and eventually, when the invasion began, against enemy ground forces. Third, German air units would attack British ports and imports; and finally, independent of the first three tasks, the Luftwaffe would launch ruthless retaliatory terror raids against major British cities (in retaliation for the present or future attacks of Bomber Command on Germany).[116]

Göring's remarks made good sense. In retrospect, Fighter Command was indeed the heart of the British defensive system. What the staff study by *Fliegerkorps I* suggests, however, is that Göring's subordinates, including his air force commanders, held other aims, which no matter how worthwhile, served to distract German strategy from the fundamental aim: destruction of Fighter Command. German decision-making during the battle reflected this confusion and the Germans proved all too willing to move from one strategy to another.

The Luftwaffe did not officially begin its offensive until mid-August with the launching of "Eagle Day" on 13 August 1940. The battle in fact began earlier than that; the British date the beginning as 10 July. The period between 10 July and 13 August saw an escalating level of Luftwaffe operations as the Germans probed their opponents over the Channel and southern British ports. The overall purpose seems to have been to wear Fighter Command down before the beginning of the main battle and to close the Channel to British maritime and naval shipping. Thus, the focal points of early air battles were the convoys along the southern coast. In retrospect, the German strategy was in serious error. It allowed the British air defense system to gain extensive experience with German operational methods. On the radar side, the British worked flaws out of the existing system, and the slow increase in the tempo of German operations gave British radar operators confidence in their abilities to estimate size and to predict the course of raids.[117] There were some major errors in the first days such as on 11 July when the radar system scrambled six Hurricanes to meet what was supposedly a lone raider making for Lyme Bay. In fact, the Hurricanes ran into a major raid of fifteen dive bombers, escorted by thirty or forty twin-engine fighters.[118] Such nasty experiences occurred with lessening frequency as the battle progressed.

The opening phase came at considerable cost to both sides in aircraft and air crews. By the second week of August, Fighter Command had lost 148 aircraft, compared to 286 for the Luftwaffe (105 Bf 109s).[119] Yet, the cost to Fighter Command in pilots was serious. In July, the loss of Spitfire and Hurricane pilots along the Channel was well in excess of the Bf 109 pilot losses in the Luftwaffe (eighty-four pilots, 10 percent, versus forty-five pilots, 4.1 percent).[120] Higher British losses, of course, reflected British tactics, which were still inferior to those used by Bf 109 pilots in air-to-air combat. Compensating for higher British fighter pilot losses was the heavy damage that British fighter squadrons inflicted on Luftwaffe bomber formations. The upshot of the preliminary phase was a standoff. The Luftwaffe forced the Royal Navy to close down the Channel convoys that had formed the focus of July air battles. But Fighter Command and its support structure had gained invaluable experience and confidence.

The weaknesses in British air-to-air fighter tactics in France and at the beginning stages of the Battle of Britain reflected the prewar dogmatism of an air staff that had argued categorically that dogfights would not take place in the next war. Consequently, RAF fighters flew in very close formations called "vics" which not only gave far less visual coverage and warning of a German fighter attack but also made it easy for Bf 109s, bouncing such a formation, to shoot down more than one of the British fighters. As combat experience spread throughout Fighter Command the British quickly adapted their tactics to fit the realities of the situation. The lesson, however, was costly.

In retrospect the prospects on the German side were less bright. Not only had the Luftwaffe tipped its hand, but nothing had yet broken the overconfidence clouding the minds of German commanders. They had in fact learned little about the workings of the British air defense system. An early August intelligence estimate announced that

> as the British fighters are controlled from the ground by R/T their
> forces are tied to their respective ground stations and are thereby
> restricted in mobility, even taking into consideration the proba-
> bility that the ground stations are partly mobile. Consequently,
> the assembly of strong fighter forces at determined points and at
> short notice is not to be expected. A massed German attack on a
> target area can therefore count on the same conditions of light
> fighter opposition as in attacks on widely scattered targets. It
> can, indeed, be assumed that considerable confusion in the de-
> fensive networks will be unavoidable during mass attacks, and
> that the effectiveness of the defenses may thereby be reduced.[121]

Thus, as planning for the assault on the British air defenses neared
fruition, the Germans had as little idea of their opponent and his tactics
as they had enjoyed at the beginning of July.

"Eagle Day (*Adlertag*)" was to begin on 10 August, but bad weather
delayed the start to the 13th. On that day the Germans again postponed
operations—to the afternoon—but too late to recall bombers, which insured that most bomber strikes in the morning had no fighter cover. As
the Germans stumbled into battle, the British felt a clear change in
tempo beginning on 11 August. On that day German fighter sweeps,
in combination with a large raid on southern ports, resulted in a furious
dogfight over the Channel that cost No. 11 Group dearly. By the end
of the day the British had lost twenty Hurricane pilots killed with two
wounded and five Spitfire pilots killed (over 7 percent of No. 11 Group's
pilots in one day). German losses were also heavy, and while losing
only twelve Bf 109 pilots, the Luftwaffe lost twenty-five other aircraft
and two more Bf 109s from which the pilots escaped unharmed.[122]

The fighting on 11 August heralded the start of massive air battles
lasting for the next week. On the afternoon of the 13th the Germans
began their attacks on the RAF and its support structure. Raids on
airfields, sector stations, and aircraft factories now became the center of
the Luftwaffe's attention. Ironically and almost inexplicably, German
intelligence misidentified the parent factory for Spitfire production in

Southampton as a bomber firm and not until much later (and for the wrong reasons) did they hit this critical target.[123] Moreover, the Germans made a serious mistake in failing to follow up their 12 August attacks on radar sites that had damaged five out of the six stations and put the Ventnor station entirely off the air until the 23rd.[124]

On 15 August, discouraged by the lack of results and the tenacity of the defenses, Göring called a meeting of senior commanders in Karinhall. While most senior commanders were absent from the battlefront, the Luftwaffe launched a series of major blows. It is doubtful whether the absence of senior commanders had much of an impact. Nevertheless, the conduct of these raids does not suggest that the Luftwaffe was absorbing and learning from its combat experience. At the same time, Fifth Air Force for the first and last time launched its aircraft against northern England in daylight and suffered a serious setback that ended its participation in the daylight offensive. The raid suffered a 15.4 percent loss to British fighters (twenty-two aircraft out of 143)—clear evidence that the British had deployed Fighter Command in depth and not in a thin shell protecting southern England. The savage air fighting on the 15th came as a terrible shock to Luftwaffe commanders, who lost seventy-five aircraft. That success had not come easy to the British—altogether Fighter Command lost twenty-six fighter pilots, killed, injured, or missing.[125]

While the British were savaging his forces, Göring, far removed at his Karinhall estate, berated his senior commanders. Not only did he criticize target selection (although failing to give substantive suggestions) but he removed radar stations from the Luftwaffe's target list.[126] Göring's decision seems to have been based partially on faulty damage estimations of what the raids on the 12th had achieved. Important to his decision would seem to have been the misestimate by Luftwaffe intelligence of the effectiveness of the British command and control system. On the following morning, German intelligence reported that heavy losses had reduced Fighter Command to 300 serviceable aircraft, but as raids over the following days met stiff opposition, doubts on intelligence estimates began to appear.[127]

Sustained bad weather beginning on the 19th brought a five-day lull. As both sides licked their wounds, the Luftwaffe's operations staff issued a new directive on 20 August. It reemphasized that the RAF, and particularly Fighter Command, was the primary target. Along with the efforts to destroy Fighter Command in the air, the air attacks should target the ground support organization, the aircraft industry, and aluminum smelting plants and rolling mills.[128] At the same time Göring finally recognized the Stuka's vulnerability to fighters and withdrew them from the battle. He also made the serious error of tying the Bf 109 fighters closely to the bomber formations. Göring's decision reflected the chorus of complaints from bomber units on the inadequacy of fighter cover and the heavy losses that bomber formations were taking from RAF attacks. It was a bad tactical mistake. Tied closely to the bomber formations, the Bf 109 force was not only less effective in its air-to-air operations against British fighters, but no more capable of protecting the bombers. In addition Göring ordered the 109s to escort the Bf 110s. Finally, the *Reichsmarschall* redeployed the single-engine fighter force, which now concentrated behind Pas de Calais under the control of Kesselring's Second Air Force.[129] While this decision provided greater support to the attacks on Fighter Command bases defending London, the decision effectively removed the Third Air Force from the daylight offensive and took the pressure off much of southern England. The pressure of sustained operations was beginning to tell on both sides. The overall losses in fighter pilots for the Luftwaffe and RAF shown in Table 7.3[130] reflect that pressure. Neither fighter force was in a position to take such losses on a sustained basis.

The bad weather ended on 24 August; three weeks of intensive operations then commenced. For the first two weeks, the Luftwaffe's target remained Fighter Command and its support structure. The Germans placed enormous pressure on the defense system. And while they did not inflict daily losses as high as they had in mid-August, they did push the British fighter forces to the limit. Luftwaffe bomber formations thoroughly devastated Fighter Command's frontline air fields and seriously stressed not just British pilots but also the command and control system and the maintenance support force. Reserves of pilots on both sides were running out. But if the British were under extraordinary pressure, the Germans saw no relaxing in the defenses. Fighter Command's resistance proved as tenacious as ever.

TABLE 7.3

Fighter Pilot Losses: Fighter Command
(Hurricane and Spitfire) and Luftwaffe (BF 109 Force)

	Fighter Command		Luftwaffe	
	Total losses all causes	percent loss fighter piilots*	Total losses all causes	percent loss of fighter pilots*
July	84	10	124**	11
August	237	26	168	15
September	264	28	229	23.1

*based on number of pilots available at the beginning of the month
** may include some late returns from the battle of France

Early in September the Germans made their final mistake. Discouraged that the current strategy did not seem to be pressuring Fighter Command sufficiently, Hitler and Göring switched the Luftwaffe's approach from an air superiority strategy to a daylight strategic bombing offensive against London. The change reflected two basic attitudes. On one hand, the German leadership was furious at the British audacity in bombing the Reich's cities. On the other hand, the Führer was undoubtedly delighted to have an opportunity to see whether ruthless "terror" bombing attacks on the "soft" British plutocracy might not lead to the collapse of the war effort. The change, of course, fit in nicely with the theories of Trenchard and Douhet that had argued that air power had negated the classic strategic lessons of history. The change did find favor in the Luftwaffe's high command. Kesselring pushed for the new strategy because he, like many in the Luftwaffe's intelligence service, believed that the RAF was on its last legs. He argued that a series of great raids on London would bring what was left of Fighter Command within reach of his fighters.

The shift in bombing strategy came with startling suddenness. On 7 September, the pressure on Fighter Command's throat entirely relaxed. Late in the afternoon, Kesselring launched 348 bombers and 617 fighters, nearly 1,000 aircraft, against London.[131] The change caught No. 11 Group so much by surprise that the response was most uneven; controllers initially reacted as if the massive raid were targeting the sector fields and the controlling network.[132] Consequently British fighters did not reach the German bombers until most had dropped their loads. In swirling dogfights south of London, Fighter Command lost twenty-two more Hurricane and Spitfire pilots, but inflicted the loss of forty aircraft, including bombers. More importantly, the Luftwaffe lost twenty-two Bf 109 pilots.[133]

The damage the Luftwaffe inflicted on London was terrible, and the respite that followed proved invaluable to hard pressed defenders, ground support, and aircrew. One week later, Kesselring's forces returned for a repeat performance. Aircrews, assured that the RAF was through, discovered what some Luftwaffe commanders had sensed at the end of August; Fighter Command was an extraordinarily resilient instrument. Air-to-air combat on the 15th worked out to a rough equivalency in fighter pilot losses, twenty British pilots to seventeen German fighter pilots, but the Hurricanes and Spitfires savaged the German bomber formations. Beside Bf 109 losses the Germans lost a further forty-one aircraft. While those losses were not catastrophic, in and of themselves, the Luftwaffe's bomber crews had reached the breaking point; many, at the first appearance of British fighters, dumped their bomb and ran for the coast.[134] Even though heavy daylight air operations continued into October, the Battle of Britain was over. The Luftwaffe had indeed failed to gain anything approximating air superiority.

The Battle of Britain was one of the most uplifting victories in human history. "The few" had indeed triumphed, but they had triumphed because of outstanding leadership on their side and sloppy, careless execution on the German side. The foremost factor in the Luftwaffe's failure lay in the overwhelming overconfidence with which it had approached the problem of defeating the RAF. That task alone represented a wholly new strategic problem, entirely beyond that with which the Germans had hitherto grappled. The cavalier attitude of an incompetent intelligence service reinforced the mood of overconfidence. Sure of its abilities and hopeful that the British would sue for peace, the Luftwaffe procrastinated from mid-June to mid-August. And in desperation the RAF, inspired by the threat to national existence, rallied its forces. Not only did the Luftwaffe dally, but by engaging in largely irrelevant operations over the Channel for nearly two months it built up the confidence as well as the expertise of Fighter Command. The badly executed "Eagle Day" was a fitting anticlimax to the confused beginning. With a commander in chief far removed from the battle, with its air fleet commanders ensconced in comfortable mansions, the Luftwaffe moved from one strategic conception to another with no clear idea of an overall strategy. But although blinded by its own intelligence as to the importance of the radar system, and misguided as to the location of fighter factories in Britain, the Luftwaffe still managed to inflict excruciating pain on Fighter Command. But that pain, without the discipline of a strategic conception, could not gain a decisive victory.

Tables 7.4,[135] 7.5,[136] and 7.6[137] show German pilot availability and losses. From the invasion of Scandinavia in April 1940 the Luftwaffe was involved in massive air operations spanning nearly all of Western Europe. Its losses, in terms of aircraft alone, were staggering. And the cumulative pressures reaching back to April finally broke the morale of some units, particularly bomber squadrons, which had been engaged more or less continuously ever since their brilliant and devastating intervention in the battles which had led to the fall of France.

TABLE 7.4

BF 109 Squadron Fighter Pilot
Availability and Losses 1940

	Pilots Operationally ready available at beginning of month		Pilots at beginning of month Pilot losses during month	
	Number of Pilots	Percent of Total	Number of Pilots	Percent of Total
May	1110	1010	91	76	6.8
June	1199	839	70	93	7.8
July	1126	906	80.5	124	11
August	1118	869	77.7	168	15
September	990	735	74.2	229	23.1

TABLE 7.5

Luftwaffe Aircraft Losses, July–September 1940
Destroyed on Operations

Type of Aircraft	Strength 29.6.40.	Due to Enemy Action	Not Due to Enemy Action	Total	Destroyed Not on Operations	Total Destroyed	Total destroyed as Percent of Initial Strength
Close recce	312	1	2	3	5	8	3
Long-range recce	257	47	14	61	9	70	27
Single-engine fighters	1,107	398	79	477	41	518	47
Twin-engine fighters	357	214	9	223	12	235	66
Bombers	1,380	424	127	551	70	621	45
Dive bombers	428	59	10	69	19	88	21
Transport	408	3	1	4	11	15	4
Coastal	233	38	29	67	14	81	35
Total	**4,482**	**1,184**	**271**	**1,455**	**181**	**1,636**	**37**

TABLE 7.6

Luftwaffe Aircraft Losses, May–September 1940
Destroyed on Operations

Type of Aircraft	Strength 4.5.40.	Due to Enemy Action	Not Due to Enemy Action	Total	Destroyed Not on Operations	Total Destroyed	Aircraft Destroyed in May–September as Percent of Initial Strength
Close recce	345	68	7	75	11	86	25
Long-range recce	321	115	32	147	11	158	49
Single-engine fighters	1,369	567	145	712	63	257	19
Twin-engine fighters	367	304	25	329	16	345	94
Bombers	1,758	862	180	1,042	100	1,142	65
Dive bombers	417	148	34	182	28	210	50
Transport	531	191	19	210	18	228	43
Coastal	241	58	45	103	17	120	50
Total	**5,349**	**2,313**	**487**	**2,800**	**264**	**3,064**	**57**

In fact, the Luftwaffe was the only air force in the world in 1940 that thought in terms of an air superiority strategy over the *enemy's territory*. Admittedly it had cast that strategy very much within a Central European *Weltanschauung*, but where Luftwaffe operations worked together with the army to remove Germany's continental neighbors that strategy was impressively effective. Air superiority, once gained by massive strikes against enemy air forces, allowed the Luftwaffe to support, to protect and to supply the army's rush into the enemy's heartland. Where the enemy did not possess the time, the resources, or the space to avoid the heart thrust, death by paralysis soon followed. With Britain the Luftwaffe faced a very different problem. The Germans did possess a strategy of air superiority, but the strategic framework of the Battle of Britain was so radically different from their experience that they never properly grasped the issues. This was particularly so since they had wasted so many assets in the waiting period of July and early August. When the Luftwaffe began its major effort in mid-August, it was already too late. British production and sage leadership were enough to keep Fighter Command in the struggle through to the period of bad weather. It is sobering to note, however, that Fighter Command's pilot losses in August and September were worse than the Luftwaffe's worst months in the January-May 1944 air battles over Germany—air battles that finally broke the Luftwaffe's back and irrevocably won air superiority over the continent for Allied air forces.[138]

There has been a condescending tendency among Anglo-American commentators on the Battle of Britain to point to the Luftwaffe's "extraordinary" mistakes. However, neither the RAF nor the American Army Air Forces possessed an air superiority strategy in 1943 in the air battles over Germany, and both paid a fearful price in terms of the lives of their aircrews. In 1944 the appearance of the P-51, almost by accident, enabled the American Air Forces in Europe to wage a successful campaign of air superiority. The immense cost of that victory underlines the price of winning and maintaining air superiority over an opponent with the resources and depth to fight an independent air war. In 1940, the Luftwaffe did not have the resources in aircrew and aircraft to wage such a battle through to victory.

NOTES

1.   Sir John Slessor, *Airpower and Armies* (London, 1936), p. 68.

2.   Public Record Office, Air 20/40, Air Staff Memorandum No. 11A, March 1924.

3.   For a fuller discussion of the development of air power theories in the interwar period, see appendix I of my study, *Strategy for Defeat, The Luftwaffe 1933-1944* (Montgomery, 1983). For other works on the development of prewar air doctrine and air power theories and their incessant themes that air war would no longer be bound by the principles of war, see Sir Charles Webster and Noble Frankland, *The Strategic Air Offensive Against Germany*, vol. I, *Preparation* (London, 1961); Anthony Verrier, *The Bomber Offensive* (London, 1968); Max Hastings, *Bomber Command* (London, 1979); John Terrain, *The Right of the Line*

(London, 1985); Thomas H. Greer, *The Development of Air Doctrine in the Army Air Arms, 1917-1941* (Montgomery, 1955); Thomas Fabyanic, "A Critique of United States Air War Planning, 1941-1944," Ph.D. dissertation, St. Louis University, 1973; Barry Watts, *Foundations of U.S. Air Doctrine: The Problem of Friction in War* (Montgomery, 1983).

4.   Sir Charles Webster and Noble Frankland, *The Strategic Air Offensive Against Germany*, vol. IV, appendix I, Minutes of a Conference Held in The Room of the Chief of the Air Staff, Air Ministry on 19 July 1923.

5.   For this view, see Dennis Richards, *The Royal Air Force, 1939-1945* (London, 1953), p. 29; Asher Lee, *The German Air Force* (New York, 1946), pp. 16-17; and even Webster and Frankland, *The Strategic Air Offensive Against Germany*, vol. I, *Preparation*, p. 125. For the weaknesses in this view, see Williamson Murray, "The *Luftwaffe* Before the Second World War: A Mission, A Strategy?," in the *Journal of Strategic Studies* (September, 1981).

6.   For the theory of Blitzkrieg strategy, see Larry Addington, *The Blitzkrieg Era and the German General Staff* (New Brunswick, N.J., 1971); Alan Milward, *The German Economy at War* (London, 1965); Burton Klein, *Germany's Economic Preparations for War* (Cambridge, Mass., 1959); and most recently F. H. Hinsley, *British Intelligence in the Second World War* (London, 1979). For a strategic analysis of why this theory is largely erroneous, see Williamson Murray, "Force Structure, Blitzkrieg Strategy and Economic Difficulties: Nazi Grand Strategy in the 1930s," *Journal of the Royal United Services Institute* (April 1983).

7.   For an outstanding discussion of how easily Hitler was able to overthrow the European diplomatic balance of power, see Gerhard Weinberg, *The Foreign Policy of Hitler's Germany*, vol. I, *1933-1936* (Chicago, 1970).

8.   For the relationship between Germany's economic difficulties and her strategic policies, see Williamson Murray, *The Change in the European Balance of Power, 1938-1939; The Path to Ruin* (Princeton, 1984), particularly chapter 1. See also Friederich Forstmeier and Hans-Erich Volkmann, eds., *Wirtschaft und Rüstung am Vorabend des Zweiten Weltkrieges* (Düsseldorf, 1975).

9.   For the best current discussion of the German rearmament effort and the general unwillingness to face the strategic, economic, and financial consequences of the rearmament programs, see Wilhelm Deist, Manfred Messerschmidt, Hans-Erich Volkmann, and Wolfram Wette, *Das Deutsche Reich und der Zweite Weltkrieg*, vol. I, *Ursachen und Voraussetzungen der deutschen Kriegspolitik* (Stuttgart, 1979). For an excellent summation of this volume in English, see Wilhelm Deist, *The Wehrmacht and German Rearmament* (London, 1981).

10.   Bernard Heimann and Joachim Scunke, "Eine geheime Denkschrift zur Luftkriegkonzeption Hitler-Deutschlands von Mai 1933," *Zeitschrift für Militärge-*

schichte, vol. III (1964).

11.   For the reappraisal of the Luftwaffe's interest in strategic bombing, see Klaus A. Maier, Horst Rohde, Bernd Stegemann, and Hans Umbreit, *Das Deutsche Reich und der Zweite Weltkrieg*, vol. II, *Die Errichtung der Hegemonie auf dem europäischen Kontinent* (Stuttgart, 1979); and also Williamson Murray, *Strategy for Defeat, The Luftwaffe 1933-1945* (Montgomery, Alabama, 1983), and Murray, "The Luftwaffe before the Second World War: A Mission, A Strategy?," *Journal of Strategic Studies* (September 1981).

12.   In 1933, Germany had no aircraft fully suitable for employment as a strategic bomber. The Ju 52 was pressed into service as an interim bomber, but the German air leadership had few illusions about its suitability for the mission. See Edward L. Homze, *Arming the Luftwaffe, The Reich Air Ministry and the German Aircraft Industry, 1919-1939* (Lincoln, 1976), chapter 9.

13.   Homze, *Arming the Luftwaffe*, p. 60.

14.   Walther Wever, "Vortrag des Generalmajors Wever bie Eröffnung der Luftkriegsakademie und Lufttechnischen Akademie in Berlin-Gatow am 1. November 1935," *Die Luftwaffe* (1936); and L.Dr.16, "Luftkriegführung," (Berlin, 1935). Copies of both of these documents were made available to me by Oberstleutnant Klaus Maier of the Militärgeschichtliches Forschungsamt, Freiburg, Federal Republic of Germany.

15.   Paragraph 18, "Die Luftkriegführung." See also paragraph 9.

16.   "Vortrag des Generalmajors Wever. . . ," p. 7.

17.   Ibid., p. 6, and paragraph 17, "Die Luftkriegführung."

18.   Paragraph 103, "Die Luftkriegführung."

19.   Air Historical Branch, Air Ministry, VII, Translations: Luftwaffe Strength and Serviceability Statistics, G302694/AR/9/51/50.

20.   Given the rapidity of expansion as well as the Germans' own inclinations, they were not able to build an organization that fully integrated the technical and engineering side of the house with the operational world. For a discussion of the defects within the Luftwaffe, see Horst Boog's outstanding article, "Higher Command and Leadership in the German Luftwaffe, 1935-1945," *Air Power and Warfare, Proceedings of the Eighth Military History Symposium, USAF Academy* (Colonel Alfred T. Hurley and Major Robert C. Ehrhart, eds.) (Washington, 1979) and his groundbreaking book, *Die Deutsche Luftwaffenführung, 1935-1945, Führungsprobleme, Spitzengliederung, Generalstabsausbildung* (Stuttgart, 1982).

21.   The extraordinary difficulty that the Germans ran into in that process are discussed in a gloomy memorandum written shortly after the Munich crisis: Der Chef des Nachschubsamts, Nr. 3365/38, g. Kdos., 3.11.38; Anlage L.7.Nr.

15.222/38, "Erfahrungsbericht über die Spannungszeit," Milch Collection, Imperial War Museum, Reel 55, vol. 57, p. 3270.

22. Karl-Heinz Volker, *Dokumente und Dokumentärfotos zur Geschichte der deutschen Luftwaffe* (Stuttgart, 1968), Doc. 184, p. 429.

23. Homze, Arming the Luftwaffe, p. 172.

24. Close air support did not receive a high priority in the Luftwaffe's allocation of resources. In fact, the tactics and command and control necessary for its implementation were largely developed in Spain by General Wolfram von Richthofen without much enthusiasm from the Air Ministry. Conversation with Generalmajor a.D. Hans W. Asmus, Baden Baden, 7 and 8 November 1980 and letter from General Asmus, 6 February 1981. For more on Spain, see "Lehren aus dem Feldzug in Spanien, Einsatz vom Schlachtfliegern," aus einer Studie der 8.Abt. des Generalstabes aus dem Jahne 1944; and Hans Herwig Freiherr von Beust, "Die deutsche Luftwaffe in spanischen Krieg," 2.10.56., p. 162, Alfred F. Simpson Historical Research Center (AFSHC): K113.302.

25. Wever's doctrinal statement, "Die Luftkriegführung," made clear that cooperation with the army could be a major role for the Luftwaffe depending on the wartime situation (see particularly paragraph 20). This, of course, was a very different attitude from that manifested by most other air forces at this time.

26. One of the persistent legends of air power history is the argument that the Luftwaffe had no interest in strategic bombing because it possessed no four-engine bombers on the outbreak of the war. In fact the Germans were hard at work attempting to make a suitable four-engine strategic bomber. See in particular Edward Homze's excellent article "The Luftwaffe's Failure to Develop a Heavy Bomber Before World War II," *Aerospace Historian* (March 1977).

27. On the lack of interest in a long range fighter in the U.S. Army Air Force, see Fabyanic, "A Critique of United States Air War Planning, 1941-1944," and Bernard Boylan, "The Development of the Long-Range Escort Fighter," unpublished manuscript (Maxwell AFB, 1955), AFSHC. The RAF attitude was quite similar; escort fighters of high performance and range could simply not be developed. See the lecture given by the future vice commander of Bomber Command R.H.N.S. Saundby to the RAF Staff College in May 1937, "Bombing Tactics," Public Record Office (PRO) AIR 5/1132, pp. 10-15.

28. Adolf Galland, *The First and the Last* (New York, 1954), p. 31.

29. Ibid., p. 63.

30. Gerhard Weinberg, *The Foreign Policy of Hitler's Germany* (Chicago, 1970), p. 298.

31. Matthew Cooper, *The German Air Force, 1933-1945, An Anatomy of Failure* (New York, 1981), p. 59.

32.   Derek Wood and Derek Dempster, *The Narrow Margin* (London, 1961), pp. 49-50.

33.   Air Ministry, *The Rise and Fall of the German Air Force, 1933-1945* (London, 1948), p. 14.

34.   For a more detailed examination of the military and strategic context within which the great powers were operating in the fall of 1938, see chapter 7 of Murray, *The Change in European Balance of Power, 1938-1939*.

35.   As in 1939 the French had no intention of carrying out any significant offensive operation against the German frontier. See particularly Gamelin's discussions with the British Chiefs of Staff in PRO CAB 21/595, 26.9.38, "Notes on a Meeting."

36.   The after action reports by the First and Third Air Forces are the basic documents for German air plans against Czechoslovakia. Both underline the *general* unpreparedness of the Luftwaffe for a military campaign against Czechoslovakia. See Bundesarchiv/Militärarchiv (BA/MA) RL7/67, Der kommandierende General und Befehlshaber der Luftwaffengruppe 1., Ia Nr 197/38, 11.7.38., Betr: "Planstudie 'Grün' 1938;" BA/MA RL 7/164, Der kommandierende General und Befehlshaber de Luftwaffengruppe 3., Ia Nr. 7829/38, 1.12.38. "Erfahrungsbericht über die Spannungzeit 1938 'Fall Grün,' Teil III;" and BA/MA RL 7/67, "Planstudie 1938, Hauptteil II, Teil A, Aufmarsch und Kampfanweisung 'Fall Grün,' zur Lw. Gruppenkommando 3., Führungsabteilung, Az Plst 38/Ia op, Nr. 525/38, 20.7.38."

37.   "Luftwaffe Strength and Serviceability Tables, August 1938-April 1945," Air Historical Branch, Translation No. VII/107.

38.   Chef des Nachschubsamts, Nr. 3365/g. Kdos. 3.11.28., Milch Collection, Imperial War Museum, reel 55, vol. 57.

39.   BA/MA RL 7/164, Der kommandierende General und Befehlshaber der Luftwaffengruppe 3., Ia Nr. 7829/38, 1.12.38. "Erfahrungsbericht über die Spannungszeit 1938 'Fall Grün,' Teil III."

40.   BA/MA RL 7/1, Der kommandierende General und Befehlshaber der Luftwaffengruppe 1., Ia Nr. 197/38, 11.7.38., Betr: "Planstudie 'Grün' 1938."

41.   BA/MA RL 7/67, "Planstudie 1938, Hauptteil II, Teil A. "Aufmarsch und Kampfanweisung 'Fall Grün,' zur Lw. Gruppenkommando 3., Führungsabteilung, Az Plst 38/Ia op, Nr. 525/38, 20.7.38."

42.   For a further examination of the military and strategic factors involved in the 1938 confrontation, see Murray, *The Change in European Balance of Power, 1938-1939*, chapter 7.

43.   Alan Bullock, Hitler: *A Study in Tyranny* (New York, 1964), p. 499.

44.   Serious problems would still show up in the offensive against Poland,

particularly in the army. See Williamson Murray, "German Response to Victory in Poland: A Case Study in Professionalism," *Armed Forces and Society*, (Winter 1981).

45. Maier, *et al.*, *Das Deutsche Reich und der Zweite Weltkrieg*, vol. II, p. 97.

46. "The Luftwaffe in Poland," a study produced by the Luftwaffe historical branch (8th Section), 11.7.44., AHB, Translation No. Y 11/33.

47. The Poles would of course show themselves to be outstanding pilots in the air defense of Great Britain in late summer 1940. For the performance of the Poles in the Battle of Britain, see Francis K. Mason, *Battle over Britain* (New York, 1969), pp. 207-8.

48. Robert Jackson, *Fighter! The Story of Air Combat, 1936-1945* (New York, 1979), pp. 27-28.

49. Air Ministry, *The Rise and Fall of the German Air Force, 1933-1945* (New York, 1983), p. 54.

50. Maier, *et al.*, *Das Deutsche Reich und der Zweite Weltkrieg*, vol. II, p. 124.

51. BA/MA, RL 3/1025, Front-Flugzeug-Verluste im September 1939.

52. For the development of these operational concepts before the war, see Air Ministry, *The Rise and Fall of the German Air Force, 1933-1945*, p. 48.

53. For a fuller examination of the strategic situation on the outbreak of the war, see chapter X of Murray, *The Change in European Balance of Power, 1938-1939*.

54. Ibid., chapters IX and X.

55. Telford Taylor, *The March of Conquest, The German Victories in Western Europe, 1940* (New York, 1958), pp. 117-18.

56. Air Ministry, *The Rise and Fall of the German Air Force*, pp. 60-61.

57. For a graphic description of the enormous difficulties under which Allied naval and land forces operated as well as the general strategic handicap resulting from German air superiority, see Martin Gilbert, *Winston Churchill*, vol. VI.

58. For those who think that the contemporary world has little need for history, one might note that the contemporary situation in the gulf states of the Middle East, with relatively ineffective air forces compared to the superpowers, invites a similar use of air power: air strikes to soften up the indigenous opposition, airborne units to seize and hold the strategic airfields, and a rapid buildup by airlift of both ground forces and ground support forces for air units (admittedly on a grander scale). Once such a force had achieved general air superiority there would be little hope of an effective counterstrike.

59. Taylor, *March of Conquest*, p. 184.

60. Air Ministry, *The Rise and Fall of the German Air Force*, p. 66.

61. See Table 7.1, for the cost to the German bombers force in the battle of France.

62.   See in particular, Patrice Buffotot and Jacques Ogier, "L'armee de l'air francoise dans la campagne de France (10 mai-25 juin 1940)," *Revue historique des Armées*, vol. II, No. 3, pp. 88-117.

63.   Major L. F. Ellis, *The War in France and Flanders, 1939-1940* (London, 1953), p. 37.

64.   Alistair Horne, *To Lose a Battle, France 1940* (Boston, 1969), p. 229.

65.   See the discussion in Galland, *The First and the Last*, pp. 2-5, which suggests that the missions that he was involved in during the first two weeks of the campaign were either screening missions to keep Allied aircraft away from German army units or direct free chase missions in which the Bf 109s aggressively sought out Allied aircraft.

66.   See discussion on the development of Luftwaffe close air support in the early war years in chapter 6 of this volume.

67.   Ellis, *The War in France and Flanders*, p. 54.

68.   "Der Einsatz der deutschen Luftwaffe während der ersten 11 Tage des Frankreichfeldzuges," Auszüge aus dem täglichen Lagemeldungen des Ober-befehlshabers der Luftwaffe, Abl. Ic., AFSHRC: k 113.306-3, v.2. The German losses are attributable to a wide variety of causes: Allied anti-aircraft artillery as well as fighters contributed to these losses.

69.   See Ellis, *The War in France and Flanders*, p. 53, for a general discussion of British losses.

70.   Horne, *To Lose a Battle*, p. 253.

71.   Ellis, *The War in France and Flanders*, pp. 55-56.

72.   Ibid., p. 56.

73.   Ibid., p. 56.

74.   "Das Jagdgeschwader 27 des VIII Flieger-Korps in Frankreichfeldzug, 1940," Generalmajor a.D. Max Ibel, 25.6.53., BA/MA, RL 10/591.

75.   Jackson, *Fighter*, p. 42.

76.   Testimony of the former chief of German air intelligence Schmid on 18.6.54., AFSHRC: K 113.306-3, v.3.

77.   Ellis, *The War in France and Flanders*, pp. 181-82.

78.   Ibid., p. 184.

79.   Einsatz des II. Fliegerkorps bei Dünkirchen am 27.5.40.: Schwerer Tag des II. Fliegerkorps," AFSHRC: K 113.306-3, v.3.

80.   Ellis, *The War in France and Flanders*, p.246.

81.   Galland, *The First and the Last*, p. 7.

82.   Heinz Guderian, *Panzer Leader* (New York, 1957), p. 75 and Maier, *et al.*, *Das Deutsche Reich und der Zweite Weltkrieg*, vol. II, p. 294.

83.   BA/MA RL 2 III/1025, gen. QU.6. Abt. (III A), "Front-Flugzeug-Ver-

luste," figures for May 1940.

84.   This table was drawn from two major compilations of the Air Historical Branch. They are AHB, Translation, VII/107, "Luftwaffe Strength and Serviceability Tables, August 1938-April 1945;" and Translation VII/83, "German Aircraft Losses, September 1939-December 1940." These tables, in turn, were compiled from the German quartermaster records then in the hands of the AHB.

85.   Two recent works have attempted to excuse the French performance in 1940 by arguing that the high command was not fully responsible due to circumstances beyond its control; see Robert J. Young, *In Command of France, French Foreign Policy and Military Planning* (Cambridge, Mass., 1978); and Jeffery Gunsburg, *Divided and Conquered, The French High Command and the Defeat of the West, 1940* (Westport, 1979). This view has been substantially refuted by the first American to secure full access to the French military documents for the 1920s and 1930s. See Robert A. Doughty, *The Seeds of Disaster: The Development of French Army Doctrine, 1919-1939,* (Hamden, Conn., 1985).

86.   Ellis, *The War in France and Flanders*, pp. 312-13.

87.   BA/MA, RL 2 III/707, Gen. Qu.6.Abt. (I), Übersicht über Soll, Istbestand, Einsatzbereitschaft, Verluste und Reserven der fliegenden Verbände. These tables were usually about five days behind actual losses. Therefore, there is some distortion, and in fact, the Germans clearly lost more pilots in May than in June (but many of the losses in May were not reported until June). What matters are the trends and the overall implication of those trends. The 15.2 percent figure suggests a heavy attrition.

88.   Murray, *The Change in the European Balance of Power, 1938-1939*, p. 273.

89.   Telford Taylor, *The Breaking Wave* (New York, 1967), pp. 53-54.

90.   Chef WFA, 30.6.40., "Die Weiterführung des Krieges gegen England," International Military Tribunal (IMT), *Trial of Major War Criminals (TMWC)*, vol. XXVIII, pp. 301-3.

91.   Jodl diary entry for 20.5.40 in IMT, *TMWC*, vol. XXVIII.

92.   For aspects of this British failure, see particularly chapters 2, 6, 8, 9, 10, and 11 of *The Change in the European Balance of Power, 1938-1939*.

93.   For a terse description of the Royal Navy's attack on its former ally see Arthur Marder, *From the Dardanelles to Oran* (London, 1974), chapter 5.

94.   Maier, *et al.*, *Das Deutsche Reich und der Zweite Weltkrieg*, vol. II, pp. 221-24.

95.   Air Ministry, *The Rise and Fall of the German Air Force, 1933-1945*, p. 75.

96.   Stephen Roskill, *The War at Sea* (London, 1954), pp. 248-49.

97.   Ibid., p. 441.

98.   Ibid., pp. 252-54.

99.   Gilbert, *Winston Churchill*, vol. VI, p. 756.

100.   Chef WFA, 30.6.40, "Die Weiterführung des Krieges gegen England," IMT, *TMWC*, XXVIII, pp. 301-3.

101.   BA/MA RL 2 II/27, "Algemeine Weisung für den Kampf der Luftwaffe gegen England," ObdL, Führungstab Ia Nr. 5835/40, 30.6.40.

102.   Table 7.2 represents a compendium of Luftwaffe strengths from Air Ministry, *The Rise and Fall of the German Air Force, 1933-1945*, p. 76, "Luftwaffe Strength and Serviceability Tables, August 1938-April 1945," Air Historical Branch, Translation No. VII/107, and Francis K. Mason, *Battle over Britain* (London, 1968), p. 186.

103.   For a translation of Schmid's intelligence estimate, see Mason, *Battle Over Britain*, pp. 612-13.

104.   Richard Overy, *The Air War 1939-1945* (London, 1980), p. 33.

105.   Mason, *Battle over Britain*, p. 130.

106.   Overy, *The Air War*, p. 33.

107.   Ibid., p. 613.

108.   Ibid., p. 613.

109.   For a devastating indictment of the Luftwaffe's intelligence service and for the basic causes of such faulty and careless work, see Boog, *Die Deutsche Luftwaffenführung*, pp. 76-124.

110.   J. R. M. Butler, *Grand Strategy*, vol. II, September 1939-June 1941, (London, 1957), p. 282.

111.   Mason, *Battle over Britain*, p. 121.

112.   This ratio is based on the figures given in Mason, *Battle over Britain*, p. 125, and the United States Strategic Bombing Survey, *Final Report*, p. 277.

113.   Based on figures in Butler, *Grand Strategy*, vol. II, p. 282.

114.   Air Ministry, *The Rise and Fall of the German Air Force*, p. 79; Basil Collier, *The Defense of the United Kingdom* (London, 1955), p. 160. For the Jodl characterization of "Sea Lion," see Chef WFA, 30.6.40, "Die Weiterführung des Krieges gegen England," IMT, *TMWC*, vol. XXVIII, pp. 301-303.

115.   BA/MA, RL 2 II/30, "Besprechung Reichsmarschall am 21.7.40."

116.   BA/MA, RL 8/1, Generalkommando I. Fliegerkorps, ABt. Ia Nr. 10260/40, 24.7.40., "Gedanken über die Führung des Luftkrieges gegen England."

117.   Collier, *The Defense of the United Kingdom*, pp. 167-71.

118.   Ibid., pp. 166-67.

119.   Ibid., p. 171.

120.   Based on the daily loss tables for July of the RAF and for the Luftwaffe available in Mason, *Battle over Britain*.

121.   Quoted in Air Ministry, *The Rise and Fall of the German Air Force*, p. 80.

122.   Percentages based on the establishment of Fighter Command on 1 August in Mason, *Battle over Britain*, p. 203.

123.   Mason, *Battle over Britain*, p. 237.

124.   Collier, *The Defense of the United Kingdom*, p. 184.

125.   Based on loss tables in Mason, *Battle over Britain*, pp. 261-64.

126.   Collier, *The Defense of the United Kingdom*, pp. 190-91.

127.   Ibid., pp. 197-98.

128.   Air Ministry, *The Rise and Fall of the German Air Force*, p. 82.

129.   Mason, *Battle over Britain*, pp. 284-85, 289.

130.   The figures in Table 7.3 for British losses are based on the combat loss tables for July, August, and September in Mason, *Battle over Britain* and on the tables he provides on Fighter Command's establishment for 1 July, 1 August, and 1 September. The Luftwaffe figures are based on the tables available in BA/MA, RL 707, 708 Gen. Qu.6.Abt. (I), Übersicht über Soll, Istbestand, Einsatzbereitschaft, Verluste und Reserven der fliegenden Verbände. The combat loss results available in Mason place the Luftwaffe fighter pilot losses at a significantly lower level: they would be July, 45 pilots, 4 percent; August, 175; 15.7 percent; and September, 177, 17.8 percent. Only the figures for July are wildly at variance and may reflect late returns from the Battle of France in the BA/MA RL 700 series.

131.   Mason, *Battle over Britain*, p. 359.

132.   Collier, *Defense of the United Kingdom*, p. 236.

133.   Mason, *Battle over Britain*, pp. 365-69.

134.   Ibid., pp. 386-95.

135.   Table 7.4 is based on the figures in BA/MA, RL 2 II/707, 708, Gen. Qu.6.Abt. (I), Übersicht über Soll, Istbestand, Einsatzbereitschaft, Verluste und Reserven der fliegenden Verbände.

136.   Table 7.5 is based on the figures in AHB, Translation VII/83, "German Aircraft Losses, September 1939-December 1940."

137.   Ibid.

138.   See Table LIV of Murray, *Luftwaffe*, p. 228. The worst month for the Luftwaffe was May 1944, when the Germans lost 25 percent of the fighter pilots on duty at the month's beginning. (Based on calculations from the following documents: BA/MA 2 III/728-731, gen. Qu. 6. Abt (I), Übersicht über Soll, Istbestand, Einsatzbereitschaft, Verluste und Reserven der fliegenden Verbände).

CLAUSEWITZ: SOME THOUGHTS
ON WHAT THE GERMANS GOT RIGHT

This chapter largely results from a discussion in fall 1984 between the author and Dr. Manfred Messerschmidt of the Militärgeschichtliches Forschungsamt (the Federal Republic's Military History Institute). In response to a suggestion that Clausewitz's influence might explain German military effectiveness at the operational and tactical levels, Dr. Messerschmidt demurred, succinctly commenting that few German officers had bothered to read Clausewitz either before or after the First World War. The direct evidence generally supports Dr. Messerschmidt. On the strategic level, as historians of German military history in the twentieth century have made clear, the German officer corps not only misunderstood but deliberately inverted Clausewitz's arguments. The depressing litany of German strategic blunders in both World Wars offers little evidence that the German military paid the least attention to Clausewitz's formulation that

> war is not a mere act of policy but a true political instrument, a
> continuation of political activity by other means. . . . War in gen-
> eral, and the commander in any specific instance, is entitled to

require that the trend and designs of policy shall not be inconsis-
tent with the means. That, of course, is no small demand; but
however much it may affect political aims in a given case, it will
never do more than modify them. The political objective is the
goal, war is the means of reaching it, and means can never be
considered in isolation from their purpose.[1]

The tragedy for Germany on the strategic level was that if the Germans read Clausewitz at all before the First World War, they read selec-
tively. As Michael Howard suggests: "But von der Goltz expressed the
majority view in refusing to accept this. It was not that he ignored the
political element in the Clausewitzian trinity. He considered it to be no
longer relevant."[2] If the Germans distorted Clausewitz's view of strategy
before the war, they gave him even less attention afterwards. The future
American general Albert C. Wedemeyer attended the Germany army's
Kriegsakademie in the mid-1930s and submitted a detailed report on his
course of instruction.[3] He makes no mention of Clausewitz, reflecting
not only his own predilections but also those of the German general
staff.

The divorce from Clausewitz's wisdom on political/military relations
began with the conflict between Moltke and Bismarck in the nineteenth-
century wars of liberation, the 1866 war against Austria, and the 1870-71
war against France. Moltke did not deny the importance of political
aims; rather he believed that when war broke out, military concerns
became the dominant player in national policy:

1. Diplomacy avails itself of war to attain its end, crucially influencing the beginnings of a war and its end. It does the latter
by reserving to itself the privilege of raising or lowering its demands in the course of the war. In the presence of uncertainty,
strategy has no choice but to strive for the highest goal attainable
with the means given. The best way in which strategy can cooperate with diplomacy is by working solely for political ends but
doing so with *complete* independence of action.

2. The course of war is predominantly governed by military considerations, while the exploitations of military success or failure
is in turn the province of diplomacy.

3. Political elements merit consideration only to the extent that they do not make demands that are militarily improper or impossible.[4]

The clashes between Moltke and Bismarck over strategic policy, particularly during the war against France, were long and harsh. Bismarck was not about to concede to the military such independence from *his* political control.[5] and in getting his own way, he helped sow the seeds of Germany's destruction. His political creation, the new German Empire, possessed no constitutional provisions to insure civilian control over the military, or even over what we would today term national security policy.

On the other side, Moltke's writings suggest confusion between strategy and operations. Indeed, Moltke may have emphasized the operational sphere of war (lying between strategy and tactics as defined by Clausewitz)[6] in German military thought in order to exclude political interference from military operations. Clearly, Moltke believed that during wartime political military concerns should be subordinate to military concerns: "in the case of tactical victory, strategy submits."[7]

In fairness to Moltke and also to Bismarck, neither man could possibly have foreseen the results of their quarrel. Of course, their quarrel did not prevent the operational success of the Prussian army, which completely altered the European strategic equation from 1866 to 1871. Unfortunately, Bismarck's successors as chancellor possessed none of his strength of character: nor did the new emperor possess the slightest inclination to overrule his generals. On the other hand, Moltke's successors were general officers who possessed little depth. If Moltke sometimes confused strategic policy and operational concerns, it is clear that his successors increasingly identified strategy with operations, denying political goals and concerns any influence over military matters, even in peace. The Schlieffen Plan, with its willful disdain for diplomatic and political realities, is the most obvious of such strategic decisions, made on the basis of pure operational necessity, which contributed to German disasters from 1918 to 1945. In casting its plans for the First World War the German general staff in 1906 addressed the potential addition of Great Britain to the anti-German coalition, following the violation of Belgian neutrality called for by the Schlieffen plan, only by predicting "that a potential British force of 100,000 men would be 'shut up' in Antwerp along with the regular Belgian army."[8]

This confusion of strategy and operations reached its zenith under Ludendorff in the First World War. The quartermaster general of the *OHL* (the army high command) defined his strategic conception for the great March 1918 German "Michael" offensive thus: "I object to the word 'operation.' We will punch a hole into [their line]. For the rest, we shall see. We also did it this way in Russia!"[9] If Ludendorff managed to lower even operational goals to the level of tactics, in his post-World War I writings he might well have changed Clausewitz's definition of the relationship between war and politics to read; peace (or politics) is a continuation of war by other means.

While these depressing facts give us insight into the causes of Germany's tragedy in both world wars, they give virtually no insight into two historical problems: how did the Germans come so close in two great wars, against virtually the rest of the world, to defeating their opponents (and they did defeat the Russians in the First World War and the French in the Second World War); and how did they hold out for so long against numerically superior forces, overwhelming economic resources, and simultaneous pressure from so many directions? There are elements in Clausewitz's writings that contributed indirectly to excellence in the tactical and operational spheres. That excellence, which marked German military performance in both world wars, helps explain Germany's perseverance.

That the twentieth-century German officer corps did not read Clausewitz is clear. In late 1949, the great British military writer, B. H. Liddell Hart, received a letter from Frhr. Geyr von Schweppenberg in which the panzer general noted that, "I have never read Clausewitz or Delbrück or Haushofer. The opinion on Clausewitz in our general staff was that of a theoretician to be read by professors."[10]

The problem in intellectual history has always been to identify the link between cause and effect. In the case of Clausewitz and the German generals, that link is at best indirect and tenuous. One can argue that Clausewitz only reflected the philosophical approach to war of the Prussian military reformer circles, on which he, as a junior member, had little effect. His only influence on the German approach to war may have been to reinforce the tactical and operational orientation of those few general staff officers who had bothered to read *On War*. On the other hand, we do know that Moltke, who had read Clausewitz carefully, was during his long tenure as chief of the general staff largely responsible for establishing the kind of training that late-nineteenth and twentieth-century general staff officers received. The conflict between those who detect substantial Clausewitzian influence on the German army's operational approach to war and those who deny such influence may in the final analysis be irrelevant. The important point is that the twentieth-century German army's approach to war on the operational and tactical planes was demonstrably Clausewitzian. The German army is perhaps the only modern army, with the possible exception of the Israeli Defense Force, to have had such an approach.

There are two areas where Clausewitz may have exercised considerable influence on those German officers who bothered to examine *On War* in detail. First, his writings reinforced the professionalism of the general staff officer that the Prussian military reformers Gneisenau and Scharnhorst had effected in response to the shattering defeats of Jena and Auerstadt. Second—an indirect, but perhaps easier-to-follow influence—his comments on operations and the battlefield (even when he is describing nineteenth-century battles) have real relevance to twentieth-century warfare. It is not surprising that Anglo-Americans, who must think first of getting to the battlefield, should concentrate on the strategic and political level of *On War*, while the Prussians and their successors, the Germans, with immediate vulnerabilities on land, would gravitate to Clausewitz's thoughts in the operational and tactical spheres.

Interestingly, Clausewitz says little about training and developing professional ethics within an officer corps. Both specifically and generally he comments on those qualities that major command in war demands. He thus devotes a whole section of Book I to a brilliant dissection of military genius but makes few specific comments on officership. Clausewitz does, however, clearly state that superior junior officers are an essential element in military performance. He tells his reader that

> in our view even junior positions of command require outstanding intellectual qualities for outstanding achievement, and since the standard rises with every step, it follows that we recognize the abilities that are needed if the second positions in an army are to be filled with distinction. Such officers may appear to be rather simple compared to the polymath scholar, the far-ranging

business executive, the statesman; but we should not dismiss the value of their practical intelligence. . . . Appropriate talent is needed at all levels if distinguished service is to be performed.[11]

Officer quality was indeed basic to the German military in both world wars. In effect, even at the most junior level the Germans preferred no officer to a bad officer.[12]

Not only were the Germans demanding in the selection process for junior officers, they also selected only a small percentage of the best to attend the *Kriegsakademie*. Then, after a demanding two-year curriculum, they honored some of those students with the general staff's crimson stripe. However, the emphasis in intellectual performance at the *Kriegsakademie* was on operational and tactical skills, while strategy fell between the cracks. The intellect and character of these future general staff officers were evaluated first by their superiors and then by their instructors at the *Kriegsakademie*.[13]

What did the Germans mean by character? Perhaps the result of the selection process suggests a definition. First and most important, during the late 1930s, the army high command possessed enough competence and intellectual honesty, albeit within a limited sphere, to recognize major flaws and weaknesses that appeared in the rearmament program and military operations. From the *Anschluss* through the French campaign, it never allowed operational success to serve as a standard of performance. No matter how stunning the Polish campaign might have appeared to outsiders, most participating units did not come up to the standards of the OKH. As has been discussed, as a result of its dissatisfaction the OKH introduced a drastic, thorough training program to correct the perceived deficiencies.[14] The German army also evaluated combat experiences and maneuvers realistically, and the higher the level of command, the more critical of performance were the officers and the higher the expectations.[15] Besides its realism and toughness, German military training incorporated past experience and managed to give the German soldier a high sense of unit identification, with a belief that he could depend on the soldier next to him. In this sense, the performance of the army in the late 1930s depended on the character of the officers who served.

The second area where Clausewitz's writings substantially are in line with the German approach is in his sense of the battlefield. His discussions on the nature of war stand in stark contrast to the dominant approach to war in the nineteenth and twentieth centuries, that of Baron Henri Jomini.[16] Jomini's analysis of eighteenth and nineteenth-century warfare attempted to reduce war to the clarity of a geometric exercise. In that exercise, much of the military thought during the twentieth century has followed Jomini's efforts to bring order to the chaos of war and to explain battles in terms of clearly elucidated principles. This phenomenon has been particularly prevalent in those services dominated by technology; the disciples of Mahan, Trenchard, Douhet, and the Air Corps Tactical School all have seen war with a clarity and precision that events at sea and in the air in both world wars have mocked. Not only practitioners but also many twentieth-century writers of military history have been Jominian, as John Keegan so clearly identified for us in the mid 1970s.[17] It is somewhat disheartening that a novel like Michael Shaara's *The Killer Angels* may tell us more about what really happened at Gettysburg than most historical accounts of that terrible battle.[18]

Such a Jominian *Weltanschauung* has had an unfortunate impact on those military services who cast their preparations for combat within such a mental framework. When military doctrine aims at a simplicity and a clarity that only the clean red and blue arrows of postwar military histories possess, it can leave the real battlefield littered with smashed up aircraft and the burnt out hulks of tanks, not to mention dead and mutilated human beings. Unfortunately, wartime commanders educated within such mental frameworks have generally attempted to make reality fit doctrinal preconceptions. The experiences of Eighth Air Force and Bomber Command in World War II are cases in point. In the former case, Eighth Air Force commanders in 1943 threw great unescorted formations of B-17s against German fighter defense until their command came close to destruction in the skies over Germany in October.[19] Similarly, Arthur Harris in 1944 nearly destroyed Bomber Command in the Battle of Berlin in his effort to prove that independent "strategic" bombing could win the war by itself.[20] D. C. T. Bennett, the commander of the Pathfinder force in Bomber Command, has suggested that the best method for avoiding such unwillingness to face reality in the upper levels of command in air forces would be to make senior air commanders fly on active operations. "For every Air Vice Marshall lost, the RAF would save 200 air crews."[21]

This Jominian mechanistic view of war has spilled over into the area of command and control. This approach, which clearly believes that a centralized command and control system is the best approach to war, is a basic theme in the military history of the twentieth century. For the most part its impact has been unfortunate whether in military operations on the Somme in 1916 or "Market Garden" in 1944.

Unfortunately for their enemies, the Germans have generally avoided the Jominian view of military operations. The German approach can only be characterized as Clausewitzian. Above all *On War* exposes the dark side of combat:

> We are not interested in generals who win victories without bloodshed. The fact that slaughter is a horrifying spectacle must make us take war more seriously, but not provide an excuse for gradually blunting our swords in the name of humanity. Sooner or later someone will come along with a sharp sword and hack off our arms.[22]

Clausewitz's depiction of war not only sets out the horror of watching limbs torn from bodies but also the psychological parameters of acting and thinking under such circumstances. In chapter 4 of Book I he describes the movement of the novice onto the battlefield. From the first terrifying sounds of battle to the "sight of men being killed and mutilated," Clausewitz sets out in a brilliant passage that has lost none of its relevance for the student of war today, the fearful impressions that assault the new recruit. "It is," he tells us, "an exceptional man who keeps his powers of quick decision intact if he has never been through this experience." And while exposure to combat may mitigate some of its impact,

> the ordinary man can never achieve a state of perfect unconcern in which his mind can work with normal flexibility. Here again we recognize that ordinary qualities are not enough. . . . Headlong, dogged, or innate courage, overmaking ambition, or long familiarity with danger—all must be present to a considerable

degree if action in this debilitating element is not to fall short of achievement that in the study would appear as nothing out of the ordinary.[23]

War then in reality has little in common with the clear, concise depictions that appear in military histories; it is blood, fear, and terror.

Above all Clausewitz saw that war, as with all human endeavors involved varying degrees of confusion, error, and incompetence. "Everything in war is very simple, but the simplest thinking is difficult." He suggests what separates real war from that envisioned by amateur strategists:

> Friction is the only concept that more or less corresponds to the factors that distinguish real war from war on paper. The military machine—the army and everything related to it—is basically very simple and therefore seems easy to manage. But we should bear in mind that none of its components is of one piece. . . . A battalion is made up of individuals, the least important of who may chance to delay things or somehow make them go wrong. The dangers inescapable from war and the physical exertions war demands can aggravate the problem to such an extent that they must be ranked among its principle causes.[24]

What Clausewitz understood so well was the immense willpower necessary to overcome the inertia of human institutions and the enormous self discipline necessary to separate the essential from the inessential. As Clausewitz's defines character, "we mean the ability to keep one's head at times of exceptional stress and violent emotion."[25] Thus, "in the dreadful presence of suffering and danger, emotion can easily overwhelm intellectual conviction, and in this psychological fog it is so hard to form clear and complete insights."[26]

There is another element in Clausewitzian thought on war to note before turning to the problem of what the Germans did absorb from his writings. That is Clausewitz's emphasis on boldness at all levels of command as an essential element in success on the battlefield.

> Let us admit that boldness in war even has its own prerogatives.
> It must be granted a certain power over and above successful
> calculations involving space, time, and magnitude of force, for
> wherever it is superior, it will take advantage of its opponent's
> weakness. In other words, it is a genuinely creative force . . .
> Whenever boldness encounters timidity, it is likely to be the win-
> ner, because timidity in itself implies a loss of equilibrium . . .
> Given the same amount of intelligence, timidity will do a
> thousand times more damage in war than audacity.[27]

Thus, in the fog, fear, and frictions of war, the victor is the individual
as well as the army that acts with calculated boldness. As Albert C.
Wedermeyer in his report to the War Department in 1938 noted, "The
Germans point out, that often a Commander must make an important
decision after only a few minutes deliberation and emphasize, that a fair
decision given in time for aggressive execution is much better than one
wholly right but too late."[28]

The problem in the twentieth century has been that success in war
has depended as much on boldness and initiative as it has depended
on material superiority. But centralized systems of command and control have *not* possessed the responsiveness or flexibility to *allow* tactical
and operational exploitation of the ever-changing situations that occur
on the battlefield. The Dardanelles campaign underlines in graphic detail that a failure to act boldly at the operational or tactical levels can
turn strategic boldness into catastrophic failure. What was indeed the
most striking strategic move of the First World War foundered on lack
of operational boldness, the unwillingness of commanders at all levels
to act independently, and the strongly centralized command structure.
Thus, the colonels at "Y" Beach awaited instructions that never came,
the Anzac troops picnicked on their way to the ridges overlooking the
Dardanelles, and General Stopford's troops organized and reorganized
themselves for seventy-two hours at Sulva Bay.[29] As Churchill so aptly
noted, "the terrible ifs accumulate." The problem in World War I was
that one can not act or react in military operations with boldness and
flexibility if one must wait to be told what to do. If that were so in

Clausewitz's day, it is even truer today with the expansion in breadth and depth of the modern battlefield.

In August 1914 the German army was no more willing to recognize this reality than were its opponents. In fact, even at the highest level the Germans displayed an aversion to delegating authority to subordinate commanders. The chief of the general staff had delegated control over the Schlieffen Plan's right wing to the commander of Second Army, General Karl von Bülow. That general, however, kept so tight a reign over the First Army and its commander, General Alexander von Kluck, that Kluck had to attack the BEF straight ahead at Mons rather than sweeping around the Allied flank.[30]

Nevertheless, in the elder Moltke the Germans did possess at the highest level something that they called *"Auftragstaktik"* — "mission oriented" tactics. What they meant by such a conception was that the overall commander set goals and objectives and provided the means but allowed his subordinates maximum flexibility in how they would accomplish their task. In both the Franco-Prussian and the Seven Weeks War, Moltke allowed subordinate army and even corps commanders great latitude in the conduct of operations. But such decentralization of control went no lower than army or corps commander. Admittedly, given the nature of the nineteenth-century European battlefield, there was no obvious need to do so.

In the early years of World War I under the uninspired leadership of General Erich von Falkenhayn, the Germans fought much as did the other European armies. By 1916 Falkenhayn's efforts to centralize control of the battlefield into his hands came close to ending Germany's war effort. At Verdun he did not allow the army commander control over his own reserves.[31] While the Germans may have achieved their strategic aim for that battle (to bleed the French white), the costs to the Germans were equally disastrous. The last stages of the Battle of Verdun saw the opening of the great British offensive along the Somme. Historians have for obvious reasons concentrated on the events of 1 July 1916, when Haig and his commanders managed to lose nearly 60,000 men. What many historians have missed is that the German army lost nearly as many men over the remainder of the battle as did the British, despite the fact that the British were attacking.[32] This loss reflected Falkenhayn's demand that his troops hold every bit of captured territory to the last man, that they man the forward edge of battlefield in great strength, and that they immediately counterattack any British gains. In their deep dugouts along the Somme, and in innumerable pointless counterattacks, German infantry took heavy casualties. The results led to the fall of Falkenhayn and his replacement by Hindenburg and Ludendorff. The latter became the key player in the dramatic alteration of German defensive doctrine in 1916-17 and of German offensive doctrine in the 1917-18 period. The new leaders of the army's high command (there was no overall high command of the war) faced national strategic bankruptcy and an army that had been exhausted on the field of battle. The strategic situation was so serious that Ludendorff spent the late summer and early fall worrying whether Denmark or Holland would enter the war on the Allied side. Ludendorff in his visits to the front discovered how serious the situation had become:

> ' At the conference in Cambrai [during Ludendorff's first trip to the Western front in 1916] those various matters [on what was happening to the front line troops] were merely touched upon. I got no more than general impressions, but there were enough to show the necessity of altering the plan of fighting and of improving the army in tactics and in equipment. On the Eastern Front we had for the most part adhered to the old tactical methods and old training which we had learned in the days of peace. Here [in the west] we met with new conditions and it was my duty to adapt to them.[33]

Ludendorff in his first visits to the western front demanded a thorough and honest briefing from the chiefs of staff of those units he examined. He expected detailed and honest reporting and not "favorable report[s] made to order."[34] Drawing on the experience of the previous two years of battle, the Germans under Ludendorff's guidance set about recasting the army's defensive doctrine.[35] One must stress that Ludendorff did not direct but rather coordinated this process and that significant input came from the chiefs of staff in the field as well as *experienced* combat officers who were serving in the front line.

By the end of 1916, the Germans had developed the first modern doctrine of defensive warfare in the era of the machine gun and artillery fire. That doctrine emphasized a lightly held forward line with successively stronger defensive positions. The heaviest concentration of reserves would lie in the rear areas to counterattack enemy penetrations. The emphasis was on flexibility and elasticity. Intellectually, it rested on the sense that Clausewitz had depicted in *On War* of the messiness of the battlefield. It consequently depended on the initiative of junior officers and battalion commanders. The Germans did not find it easy either to conceptualize or to implement the new defensive doctrine. Interestingly, the senior defensive expert on the western front, Fritz von Lossberg, had serious doubts about the new doctrine as being too liberal in allowing troops in the front line to pull back, but in implementing the new doctrine in Flanders in 1917, he was won over.[36]

From the basis of this defensive doctrine and their experience in the counterattacks to support its conception of elastic defense, the Germans moved in 1917 to solve the offensive problem of the First World War—how to get the soldier not just across but *through* the enemy defensive lines. This doctrine depended on the willingness of officers *at all levels* to act independently without waiting to be told what to do. It decentralized decisionmaking to the lowest practicable level and demanded that officers at all levels *react* according to the situation and not *await* orders. Above all, the new offensive doctrine demanded boldness and movement forward. In fact, it reflected the reality of the modern battlefield, where communications disappeared in a maelstrom of artillery fire, where soldiers fight, advance, and die in small groups, and where opportunities occur for only fleeting moments in time. As Michael Howard has pointed out in the Princeton edition of Clausewitz's *On War*,

> in the enormous armies of 1900, their communications dependent at best on fragile field-telephones, their size and complexity rendering elaborate maneuver out of the question, commanders in chief could give only the broadest of directions to their subordinates and rely on their intelligence and initiative to carry them out in detail. Junior officers were likely to find themselves isolated on vast battlefields in a strange, sometimes barely endurable environment, with no recourse save their inner strength to keep them going and their common sense to tell them what to do.[37]

The German army was the only military organization to recognize this fact and by the end of the war had developed a doctrine for both offensive and defensive warfare that institutionalized decisionmaking on the common sense, tactical training, and leadership qualities of frontline officers.

What the Germans meant by doctrine was fundamentally different from that of most other military organizations. Above all the German army aimed at providing a *general* tactical and operational framework to guide battlefield commanders. Clausewitz's description of what theories of war should aim to achieve in educating the minds of commanders comes as close as one can come to describing the role that doctrine played in the German army. He suggests that

> it is only analytically that these attempts to theory can be called advances in the realm of truth; synthetically, in the rules and regulations they offer, they are absolutely useless. They aim at fixed values; but in war everything is uncertain, and calculations have to be made with variable quantities. They direct the inquiry exclusively toward physical quantities, whereas all military action is intertwined with psychological forces and effects. They consider only unilateral action, whereas war consists of a continuous interaction of opposites.[38]

Thus, doctrine should be descriptive rather than prescriptive. It should provide a general framework within which one can ask the right questions. It should not aim to provide the commander with *answers*.

> [Theory] is an analytical investigation leading to a close *acquaintance* with the subject; applied to experience . . . . it leads to thorough *familiarity* with it. The closer it comes to that goal, the more it proceeds from the objective form of a science to the subjective form of a skill, the more effective it will prove. . .[39]

What these doctrinal changes in the German army attempted to do in 1916 and 1917 was to give frontline leadership a more realistic framework within which to conduct offensive and defensive battles. As Timothy Lupfer has suggested in his evaluation of the army's First World War doctrine:

> German doctrine achieved the balance between the demands of precision for unity of effort and the demands of flexibility for decentralized application. With clearly stated principles, the doctrine provided thorough, consistent guidance for the training, equipping, and organizing of the army. However, this consistency was not rigid, for in the battlefield application, the doctrine provided sufficient flexibility to accommodate the demands of local conditions and the judgement of several commanders.[40]

Above all the doctrine aimed to provide room for refinement as battlefield experience suggested changes and new avenues of approach. The devastating nature of the German successes in the spring 1918 offensives show the degree of superiority that the Germans achieved on the basis of its new doctrine.[41] Yet that operational success also underlines the incapacity and failures of the Germans at the strategic level. In spring 1918, Ludendorff aimed to punch holes in the Allied lines with the strategic purpose of seeing what might turn up. Considering the fact that the German army was already scraping the bottom of the manpower barrel, that it could barely assemble forty-plus attack divisions on the western front for its first offensive in March (out of 200 divisions), and that American forces were already beginning to arrive in substantial number, the whole design, to punch holes in the Allied lines, lacked any strategic reality.

What is interesting in the preparations for the spring 1918 offensive was the ability of the German army, having only in fall 1917 established the new offensive doctrine, to implement training on a consistent and coherent basis throughout those attack divisions that would launch the coming attacks.[42] This was done with a massive, intensive, and thorough program that schooled divisional officers and then worked up the attack divisions for the great western offensive. All training was done on the basis of the new offensive doctrine: "Attack in Position Warfare (*Angriff im Stellungskrieg*)." The ability of the Germans to perform in skilled fashion on the battlefields of both world wars rested on this close relation between doctrine and training.

The army's training, as Clausewitz argued, aimed at placing maximum pressure on troops and officers. Clausewitz suggests that there are only two sources of military spirit and they must interact in order to create it:

> The first is a series of victorious wars, the second, frequent exertions of the army to the utmost limits of its strength. Nothing else will show a soldier the full extent of his capabilities. The more a general is accustomed to place heavy demands on his soldiers, the more he can depend on their response. A soldier is just as proud of the hardships he has overcome as of the dangers he has faced.[43]

In another passage, Clausewitz estimates how important vigorous training is to the preparation of even the common soldier for combat.

> It is immensely important that no soldier, whatever his rank, should wait for war to expose him to those aspects of active service that amaze and confuse him when he first comes across them. If he has met them even once before, they will begin to be familiar to him. This is true even of physical effort. Exertions must be practical, and the mind must be made even more familiar with them than the body. When exceptional efforts are required of him in war, the recruit is apt to think that they result from mistakes, miscalculations, and confusions at the top. In consequences, his morale is doubly depressed. If maneuvers prepare him for [such] exertions, this will not occur.[44]

But there is another element involved in training an army and that is the intellectual preparation of its officer corps for war. Not surprisingly Clausewitz admits that nothing can really prepare an officer corps for the reality of combat. "Peacetime maneuvers," he suggests:

> are a feeble substitute for the real thing; but even they can give an army an advantage over others whose training is confined to routine, mechanical drill. To plan maneuvers so that some of the elements of friction are involved, which will train officers' judgement, common sense, and resolution is far more worthwhile than inexperienced people might think.[45]

The training of general staff officers in the post-World War I era suggests why Clausewitz's *Weltanschauung* was so important and useful to the Germans.[46] We know a good deal about the nature of the *Truppenamt's* training program and that of the *Kriegsakademie* after it was resurrected during the first years of Hitler's regime. Despite the prohibitions of the Treaty of Versailles, the post-World War I German army was taken over lock, stock, and barrel by the general staff. General Hans von Seeckt played a critical role in the recreation of the German army after the troubles of postwar demobilization, revolution, political turmoil, and the severe limitations on the army's size imposed by Versailles. Seeckt insured that the general staff would dominate the new army in a fashion that it had not done even during the last war.[47] He not only brought over from the old Imperial Army the whole general staff, but insured that they held the command positions—thus the general staff controlled the command *and* the staff functions in the new Reichswehr. There were a few outstanding on-general staff combat officers who also came over to the army, but few rose to positions of authority in the late 1930s and early 1940s. Erwin Rommel was the most notable example of this type of officer, and the uniqueness of his rise to a top command position in World War II is suggested by the repetitive notations in postwar memoirs that Rommel did not possess a general staff background.

General staff officers in the new *Reichswehr*, men like Leeb, Manstein, and Wever, had all played important roles in the development of German defensive and offensive doctrine on the western front at the end of World War I. They brought with them a sense of decentralized, aggressive military operations that the late war doctrine had crystallized. Their position and the domination of the general staff within the new army insured that the army would remember the realities of infantry and artillery combat on the western front and not retreat into the happy ignorance of "the real business of soldiering." Consequently, postwar German doctrine emphasized decentralized, rapid decision-making, as well as other factors such as swift exploitation, surprise, the reinforcing of success, and the masking of enemy strong points. In a real sense the *Reichswehr* was the only military force that consistently recognized the complexity of coordinating firepower and maneuver on the modern battlefield. The entire general staff and its subordinates in the rest of the officer corps absorbed those lessons, not just a few enthusiasts of armored warfare. That explains how so many German officers in the infantry and artillery branches were able to move on to armored, mechanized warfare with ease in the late 1930s, while officers from similar backgrounds in other armies could not make the transition. German officers understood the principles of exploitation; others did not.

The training of general staff officers in the interwar period insured that new officers, moving up in the army, who no longer possessed combat experience, absorbed this approach to war. One can assume that when Clausewitz's *On War* appeared in the curriculum of the *Kriegsakademie,* students as well as instructors instinctively gravitated to the sections that dealt with the psychological aspects of combat, the frictions of war, the qualities demanded of great generals, and the necessity for swift, ruthless boldness. We can also assume that the passages dealing with strategy, the supremacy of policy over military considerations found virtually no audience at all. The training itself with numerous general staff rides, field maneuvers with and without troops, and what today would be termed command post exercises, took up most of the students' time. The students studied considerable amounts of military history, but even this portion of their studies emphasized the curriculum's operational concerns.

The interest in hands-on learning suggests what the *Kriegsakademie* wished to achieve in educating its students. The general staff rides presented students with a wide variety of tactical and operational problems and expected them to react to described situations in a short period of time. The officers were not expected to come up with textbook solutions.

Nor did the instructors in the *Kriegsakademie* aim to achieve unanimity of thought. Only in a general sense were future general staff officers taught to think in the same fashion. Rather than looking for how closely students came to a "staff college solution," the instructors evaluated performance on boldness, flexibility, capacity to adapt, and steadfastness. As one recent historian of the general staff during the Reichswehr has noted:

> Because General Staff instructors believed that textbook instruction alone could not achieve the objective of flexible response, training stressed certain personal attributes—responsibility, imagination, initiative—which normally did not appear in the course syllabus and instructional materials. Candidates were expected to accept responsibility for their actions willingly, to use their imaginations, and to take the initiative. Instructors always challenged the officers to apply their knowledge to practical situations. Clearly trainees were also required to possess a certain kind of stamina because the program was such that only officers who were willing and able to spend long and tedious hours mastering a variety of difficult subjects could succeed. . . .
>
> Yet how was it possible to develop the characteristics of responsibility, imagination, and initiative in a meaningful, practical way? The answer was never very clear. More often than not the leaders resorted to stressing the even more elusive quality of character, as if it could produce flexible response to tactical problems. As future standard bearers of a proud tradition, General Staff candidates were expected to demonstrate integrity and impeccable personal behavior above all other attributes. Although the third year curriculum included a subject dealing with General Staff duties, there was no formal course on character building as such because character development was expected to result from the personal examples set by the instructors, . . . and the extremely high standards demanded in course work and personal behavior.[48]

Captain Wedemeyer's final report on his experiences at the *Kriegs-*

akademie in the mid-1930s indicates that the same approach was still in place during his term in Berlin. Student papers on tactical problems did not receive grades: rather they were returned with comments. His instructors made clear that four general points framed any evaluation of student work: "1. *Logical combat thinking.* 2. *Good tactical judgement.* 3. *Timely and aggressive application of principles.* 4. *Seizure and maintenance of the initiative.*" The instruction placed "a premium on quick logical thinking and decisions, transmitted to the command in clear concise orders for aggressive coordinated action."[49] The curriculum and the training exercises also aimed to inculcate future general staff officers with the need to act boldly and decisively. As Wedemeyer characterized this emphasis: "Better a faulty plan or decision permeated with boldness, daring, and decisiveness, than a perfect plan enmeshed in uncertainty."[50]

The *Kriegsakademie* also hoped to achieve a high level of conciseness and clarity in the orders written by future general staff officers. Throughout history there have, of course, been generals who possessed a natural ability to express their intentions with clarity.[51] But military history is also replete with commanders who could not express themselves with such clarity and conciseness. A substantial portion of the training aimed at developing these abilities in students. Again as Wedemeyer put it in his final report: "Invaluable experience, in combat exercises and maneuvers . . . contribute to a professional knowledge, that renders cooperative initiative almost axiomatic."[52]

The German army that invaded Poland in September 1939 reflected the strengths as well as the weaknesses of its leaders. On the tactical and operational levels it was a flexible, adaptive, and bold instrument of national power.[53] The army high command's reaction to weaknesses that army units displayed suggests an extraordinarily high level of expectations; and the preparations and execution of the invasion of France and the Low Countries in May 1940 show how quickly the Wehrmacht could absorb the lessons of combat and then apply them with ruthless boldness in the next battles.

And yet the German army's performance in World War II reflected its ambivalence towards Clausewitz. On one hand, its training of officers and enlisted men, its emphasis on decentralized decisionmaking, its belief that even a bold error was preferable to delay, its sense of the psychological pressures as well as the fog of war made its *Weltanschauung* truly Clausewitzian. Its opponents, for the most part, with their rigid emphasis on carefully controlled operations and careful, cautious preparations were clearly Jominian. The contrast in terms of effectiveness could not have been more devastating in terms of the battlefield results of the 1939-41 period.

On the other hand, there were few German officers who had any sense of the strategic and political analysis that marked Clausewitz's writings. A few senior officers did see the strategic issues with some clarity, the chief of the general staff from 1934 to 1938, Generaloberst Ludwig Beck in particular. Beck's memoranda, written in the spring and early summer of 1938 against Hitler's aggressive and risky[54] policy of confrontation, stand as a monument to clearheaded strategic analysis. Unfortunately, they did not represent the opinions of even a small percentage of senior generals. With Beck's resignation at the height of the Czech crisis in August 1938, the army and its generals sank back into a confusion of strategy with operations and allowed Hitler to monopolize what they called the political realm. Nothing shows this more clearly than a July 1938 letter Beck received from a bright young general staff product, General Erich von Manstein (Beck's protégé). Manstein would soon in World War II prove himself not only to possess one of the best staff minds in the army, but to be one of the army's foremost battlefield commanders. Manstein wrote to Beck to persuade him not to resign. In a long letter he argued that Beck's presence insured that the army high command, the OKH, would continue to dominate strategy and that Beck's presence as chief of the general staff prevented the armed forces high command (the OKW) from achieving a significant position in strategic decisionmaking. Only parenthetically in mid-letter did Manstein turn to the subject that was in fact forcing Beck out of the army: the strategic aspects surrounding Hitler's confrontation with Europe over the Sudetenland. Hitler, Manstein suggested, "possessed the final responsibility for national policy and had he not thus far judged the political situation correctly?"[55]

The tragedy for Germany was that such thinking all too often passed for strategy. Such dilettantism helped launch the Second and Third Reichs on two great world wars that Germany had little prospect of winning. Unfortunately, the tactical and operational skills of the German officer corps then insured that both wars would last to the point where they not only destroyed Germany, but came close to destroying western civilization. Clausewitz would not have been pleased with what his pupils had wrought.

NOTES

1.   Carl von Clausewitz, *On War*, ed. and translated by Michael Howard and Peter Paret (Princeton, 1976), p. 87.

2.   Michael Howard, "The Influence of Clausewitz," p. 32.

3.   Final report of Captain Albert C. Wedemeyer on his tour at the Kriegsakademie, "German General Staff School," Report 15,999 dated 7-11-38 from the Military Attaché, Berlin, 1Kb 6/23/39, National Archives.

4.   Gerhard Ritter, *The Sword and the Scepter, The Problem of Militarism in Germany*, vol. I, *The Prussian Tradition*, 1740-1890 (Coral Gables, Fla., 1969), pp. 194-95.

5.   For a concise, clear formulation of the clash of wills between Bismarck and the generals, see Gordon A. Craig, *The Politics of the Prussian Army, 1640-1945* (New York, 1964), pp. 202-16.

6.   Clausewitz's definition of tactics and strategy: "According to our classification then, tactics teaches *the use of armed forces in the engagement* [thus combining what we would call the operational and tactical spheres]; strategy, *the use of engagements for the object of war*." Clausewitz, *On War*, p. 128.

7.   Hajo Holborn, "Moltke and Schlieffen: The Prussian-German School," in *Makers of Modern Strategy, Military Thought from Machiavelli to Hitler* (Princeton, 1973), p. 180.

8.   Holger Herwig, "The Dynamics of Necessity: German Military Policy During the Great War," in *Military Effectiveness*, vol. I, edited by Allan R. Millett and Williamson Murray (London, 1988), p. 87.

9.   Ibid., p. 99, quoted from Crown Prince Rupprecht of Bavaria, *Mein Kriegstagebuch*, ed. by Eugene von Frauenholz, vol. II (Munich, 1929), pp. 322 and 372 note.

10.   Letter from Leo Geyer von Schweppenberg to B. H. Liddell Hart, 3.8.49, Liddell Hart Papers 9/24/61, Kings' College Library, London.

11.   Clausewitz, *On War*, p. 111.

12.   For an excellent examination of the German approach to officership in World War II as contrasted to the American approach, see Martin van Creveld, *Fighting Power: German and U.S. Army Performance, 1939-1945* (Westport, Conn., 1982).

13.   For a discussion of this emphasis on character, see David N. Spires, *Image and Reality, The Making of the German Officer* (Westport, Conn., 1984).

14.   See chapter 10.

15.   This observation comes from my extensive research in the German military documents of the late 1930s. As one studies the Erfahrungsberichte (after action reports) of the Anschluss, the Sudeten crisis, or the invasion of Poland, one sees a more and more critical examination of what had happened the higher

the level of the headquarters. This stands in *stark* contrast to this author's experience in the United States Air Force in the late 1960s. Nothing in my experience as a reserve officer, commentator on military doctrine, or graduate instructor of serving American military officers suggests to me that *anything* has changed.

16.   See particularly Baron Henri Jomini, *The Art of War*, translated from the French by G. H. Mendell and W. P. C. Craighill (Philadelphia, 1862).

17.   See John Keegan, *The Face of Battle* (New York, 1977), chapter 1.

18.   Michael Shaara, *The Killer Angels* (New York, 1974).

19.   For the most thorough examination of the attempt of Eighth Air Force Commanders to prove doctrine correct at terrible cost to their crews see Barry D. Watts, *The Foundations of U.S. Air Doctrine, The Problem of Friction in War* (Maxwell AFB, 1985); and Thomas A. Fabyanic, "A Critique of U.S. Air War Planning, 1941-1944," (Ph.D. dissertation, St. Louis University, 1973).

20.   For the most detailed examination of Bomber Command's operations, see Sir Charles Webster and Noble Frankland, *The Strategic Air Offensive Against Germany* (London, 1962). On the realities that faced the bomber crews in the Battle of Berlin, see Max Hastings, *Bomber Command* (London, 1979).

21.   Taped interview with D. C. T. Bennett, RAF Staff College Library, Bracknell.

22.   Clausewitz, *On War*, p. 260.

23.   Ibid., p. 114.

24.   Ibid., p. 119.

25.   Ibid., p. 105.

26.   Ibid., p. 108.

27.   Ibid., pp. 190-91.

28.   Final report of Captain Albert C. Wedemeyer, p. 12.

29.   For the best description of the Gallipoli campaign, see Alan Moorehead, *Gallipoli* (New York, 1956).

30.   Bernadotte E. Schmitt and Harold C. Vedeler, *The World in the Crucible, 1914-1919* (New York, 1984), pp. 43-44.

31.   For the best discussion of the Battle of Verdun, see Alistair Horne, *The Price of Glory: Verdun 1916* (New York, 1962).

32.   For an excellent description of the battle on the Somme, see A. H. Farrar-Hockley, *The Somme* (London, 1964).

33.   Erich von Ludendorff, *Ludendorff's Own Story, August 1914-November 1918*, vol. I (New York, 1919), p. 324.

34.   Ibid., p. 24.

35.   There are two first class works on this processes, Timothy Lupfer, *The Dynamics of Doctrine: The Changes in German Tactical Doctrine During the First World War* (Leavenworth, 1981); and G. C. Wyne, *If Germany Attacks, The Battle of Depth in the West* (Westport, Conn., 1976).

36.   Lupfer, *The Dynamics of Doctrine,* p. 22.

37.   Howard, "The Influence of Clausewitz," pp. 34-35.

38.   Clausewitz, *On War,* p. 136.

39.   Ibid., p. 141.

40.   Lupfer, *The Dynamics of Doctrine,* p. 55.

41.   For an excellent depiction of the first of these offensives, see Martin Middlebrook, *The Kaiser's Battle, 21 March 1918: The First Day of the German Spring Offensive* (London, 1978).

42.   Reichsarchiv, *Der Weltkrieg, 1914 bis 1918,* vol. XIV, *Die Kriegführung an der Westfront im Jahre 1918* (Berlin, 1944), pp. 41-42; see also Ludendorff, *Ludendorff's Own Story,* pp. 200-211.

43.   Clausewitz, *On War,* p. 189.

44.   Ibid., p. 189.

45.   Clausewitz, *On War,* p. 122.

46.   For the training of the general staff during the period of the Weimar Republic, see Spires, *Image and Reality, The Making of the German Officer;* for the training during the mid 1930s the Wedemeyer report provides a useful, though at times limited, view of what the instructors and curriculum at the Kriegsakademie were trying to achieve.

47.   Symptomatic of considerable constraints on the authority of the general staff and even of Ludendorff is the fact that they were barely able to persuade the War Ministry in 1918 to promote the army's *chief* artillery expert, Georg von Brückmüller, from lieutenant colonel to colonel. Bradley J. Meyer, "Innovation and Expertise: Some Changes in German Tactical Doctrine During World War I," (MA thesis: The Ohio State University, 1981), p. 81.

48.   Spires, *Image and Reality,* pp. 47-48.

49.   Final Report of Captain Albert C. Wedemeyer, p. 15.

50.   Ibid., p. 18.

51.   For an example of this natural gift, Grant's strategic orders to Sherman (4 April 1864) and to Meade (9 April 1864) are models for such clarity. See U. S. Grant, *Personal Memoirs of U. S. Grant,* vol. II (New York, 1886), pp. 130-32, 134-37.

52.   Final Report of Captain Albert C. Wedemeyer, p. 78.

53.   As chapter 10, "The German Response to Victory in Poland: A Case Study in Professionalism," suggests there were of course serious weaknesses in the army that invaded Poland. For a graphic description of those weaknesses consult Heinz Guderian, *Panzer Leader* (New York, 1961), pp. 50-51.

54.   For a discussion of the military balance at the time of Munich, see Murray, *The Change in the European Balance of Power,* chapter 7.

55.   Bundesarchiv/Militärchiv, Freiburg, N28/3, Kommandeur der 18. Division, Liegnitz, 21.vii.1938.

CHAPTER NINE

FORCE STRUCTURE,
BLITZKRIEG STRATEGY
AND ECONOMIC DIFFICULTIES:
NAZI GRAND STRATEGY IN THE 1930s

The interrelationships between force structure, military planning, economic realities and political goals are the critical determinants in how realistically a nation casts its national security policy on the international stage. This is as true for Nazi Germany as for any other power. Unfortunately, the relationships between these factors are not quantifiable and are by nature indefinite, obscure, and complex. Some Anglo-American historians have developed a theory of Blitzkrieg strategy that has gained widespread acceptance in academic circles.[1] This theory posits a German grand strategy that combines a deliberate decision on the part of Hitler to maintain consumer production at a high level, thus limiting military production, while at the same time preparing for eventual conflicts in which German tanks and aircraft working in close cooperation would wage a series of short mobile wars with pauses between campaigns to rebuild stockpiles. The theory is a masterpiece of historical ingenuity, building a clear conceptual framework around the facts.

However there is one major defect: it is wrong. The first problem is that historians, in a desire to inject order and clarity into a confused and disorderly process, recognized relationships between certain factors and policy decisions that simply did not exist. The second problem is that Anglo-American historians have chosen the American economy as their model of comparison to the German economic situation. Unfortunately, the lessons drawn from American economic performance in World War II have little in common with the economic problems that Nazi Germany faced in its rearmament effort before the outbreak of war. Furthermore, an analysis of certain prewar indicators such as aircraft and tank production, all at surprisingly low levels, if one sees a relationship between the development of modern mechanized tactics and Hitler's supposed Blitzkrieg strategy, can create a most misleading impression. This chapter examines certain aspects of Germany's economic situation, rearmament, and foreign policy in order to give a more realistic picture of German grand strategy in the last years of peace.

From the outset, the Germans faced considerable problems with their rearmament efforts. The economic capacity for a rapid and substantial increase in weapons production simply did not exist and could not be created overnight. A single example suffices to underline the nature of the initial rearmament problem. In January 1933 the entire work force of the German aircraft industry consisted of 4,000 men.[2] No possibility of an increase in aircraft production existed until a massive infusion of trained, skilled workers had occurred, coupled with a similar extension of manufacturing capacity. A second and more important factor affecting German rearmament in the late 1930s drastically impacted on weapons production and foreign policy. As a resource-poor nation, Germany had to export finished goods to pay for raw material imports required by both rearmament and export industries. In the depression years of the 1930s Germany found it difficult to find export markets to offset the cost of raw materials needed by expanding rearmament industries. As a result, holdings of hard currency shrank to zero and for most of the 1930s the Germans lived a hand-to-mouth existence in terms of their imports.[3]

Such import difficulties in no way limited either Hitler's demands for further rearmament or his extravagant long-range goals.[4] Nevertheless, beginning in 1937 a series of increasingly grave economic crises shook the fragile German economy and provided impetus to Hitler's confrontational diplomacy of 1938 and 1939. The worst of these crises came in late autumn 1938. Göring, in his capacity as head of the Four Year Plan, admitted the exacerbation of economic difficulties by the exhaustion of the labor pool, lack of any additional factory capacity, and the complete depletion of foreign exchange reserves.[5] The result of these economic troubles forced the regime early in 1939 to reduce Wehrmacht steel allocations 30 percent, copper 20 percent, aluminum 47 percent, rubber 30 percent, and cement 25-45 percent.[6]

More than just economic constraints hampered the Nazi regime. Burton Klein suggests that the Nazi regime might have invested more in military preparations for war and less on civilian consumer goods production.[7] But the Nazi regime's popularity and Hitler's successes as a dictator in the 1930s rested as much on creating a modicum of economic well being for the German people as on his foreign policy triumphs. Thus, both political and military constraints limited what Hitler could do with his rearmament program.

Nevertheless, despite these limitations, German rearmament in the 1930s was significantly greater than that of its European neighbors. The regime kept down consumer production at the expense of capital goods production[8] and concentrated the bulk of the work force in the production of capital goods and armaments. In the rearmament sector the Germans outspent their neighbors by a wide margin. In the 1935-39 period Germany spent 12.9 percent of GNP on armaments with a high of 16.4 percent in 1938. The British average for the same period was 5.5 percent with a high in 1938 of 8.3 percent. Surprisingly, the French aggregate reached only 6.9 percent with a high of 8.5 percent in 1938. Even more sobering is a direct comparison of the actual monetary value of rearmament. The Germans spent $18,379,000,000 in the four year period 1935-38, while the British and French together only managed to spend $8,780,000,000.[9] As a result the German rearmament effort proved massive enough in comparison to those of its competitors to provide the *Reich* with a narrow margin of superiority on the outbreak of the war.

After the war, the OKW economic expert, General Georg Thomas, pointed out that Germany had rearmed in breadth rather than depth.[10] However Thomas's implication that there were other options available is misleading. What must be stressed is that few Germans had any clear idea of how the Reich could escape these economic difficulties.[11] Hitler consistently dismissed the nature of the economic problem and believed that the key to Germany's eventual victory lay in the racial and ideological superiority of its soldiers. Conversely, most generals, while recognizing the economic problems, saw no tactical or operational answer to the strategic problem.

When Hitler came to power in January 1933, he made clear to his generals that his regime had no intention of settling for anything less than domination of the European continent from the Urals to the Bay of Biscay.[12] From 1923 on Hitler underlined that he did not aim at a restoration of the pre-1914 Reich, but rather at a German state that ruled all of Europe. We may doubt how seriously the generals took Hitler's talk of conquest on a continental scale, but not their enthusiasm for the new regime which granted them the financial and political support for a massive rearmament effort. From the beginning a divergence of views existed between Hitler, who aimed at what can at best be described as inordinate goals, and a military who though certainly attracted by the Nazi goals, remained too hardheaded and realistic to believe that such goals were realizable in their lifetime. Content to regain Germany's pre-eminent position on the continent, they had no clear ideas on how to win the next world war, as their reactions to Hitler's foreign policy adventures in 1938 suggests.

If the generals had no idea of the force structure required to defeat the major European powers, neither did Hitler. It is on the military side of the argument that the Blitzkrieg strategy theory collapses. First of all, Hitler played virtually no role through 1938 in casting German rearmament or in the evolution of German doctrine and strategy.[13] On the army side, the generals in charge of expansion emphasized a conventional army that in terms of its force structure differed only slightly from the French army across the Rhine. Consequently, the emphasis in prewar German rearmament remained on World War I-style infantry divisions and not on the mechanized forces that B. H. Liddell Hart and other theorists had been advocating in Great Britain. German infantry divisions marched on foot supported by horse drawn supplies and artillery.[14] In terms of resources, German armored forces received no more than their French equivalents.[15] Unlike the French, however, the German high command allowed German armor advocates greater freedom in how they developed and trained their forces; the French allowed no such freedom. Though Hitler showed some interest in tank forces, he does not seem to have done anything much to help them nor did he possess any real understanding of what modern, mechanized exploitation tactics involved.

Air rearmament also poses major challenges to the Blitzkrieg theory. The generally held view has been that the Luftwaffe designed its force structure to support the army in the lightning war campaigns that Hitler waged between 1939 and 1941. Historians have ascribed the Luftwaffe's ultimate failure to its disregard of the importance of "strategic" bombing for the coming war. In fact, the documents do not support such a contention. Interestingly, this picture itself originated from the studies done by former Luftwaffe officers after World War II who undoubtedly hoped to ease their own spotty national record by eliminating "strategic" bombing from the Luftwaffe's prewar preparations. Unfortunately, virtually every air force article in the major German military journals of the 1930s stressed "strategic" bombing as the *raison d'être* of the Luftwaffe.[16] As in other air forces of the time, Luftwaffe officers were intrigued with the possibility of "strategic" bombing in all its various aspects.

At the conclusion of a spring 1939 war game assessing the possibility of attacking Britain, General Helmuth Felmy, commander of *Luftflotte 2*, commented to his listeners that while his forces lacked the strength to achieve a military success in "strategic" bombing, the panic that had broken out in September 1938 suggested that an all-out attack on London might cause British morale to collapse.[17] The Luftwaffe did not live up to such high prewar expectations because, like the airmen of other nations, German air force officers considerably over-estimated the effectiveness of their weapon. They did not possess the force structure or the aircraft capability to conduct "strategic" air war because no one could foresee what was required. But like their counterparts in the United States and Great Britain, they envisioned the use of their air force in just such a mission and firmly believed that "strategic" bombing represented the premier mission of an air force.

The Luftwaffe figured highly in Hitler's prewar calculations of military strength. Over 40 percent of the defense budget went to the new service and there is no doubt that the Luftwaffe played a major role in Hitler's decision to go to war in 1939.[18] Interestingly, only a small percentage of the Luftwaffe's budget went into the support of the army mission.[19] Of even greater interest is the fact that the close air support doctrine for ground forces developed in an even more haphazard fashion than did that of the army's armored forces, and received little support from the Luftwaffe high command. In other words, no one in the Luftwaffe high command saw the need for such a capability, but rather those serving in the Condor Legion in Spain developed the doctrine and tactics of close air support to meet local military circumstances.[20]

What then is left in terms of Nazi rearmament, military force structure and grand strategy? To begin with, Hitler had a clear sense of his long-range goals and of the Europe he wished to create. He fully recognized the lofty nature of his aims and at least through 1939, the difficulties that he faced. He had a sense of fighting against the tide of history and, as he suggested to his generals shortly before the attack on Poland, he considered himself a historically unique individual who alone could realize the new Germany.[21] Upon his accession to power in 1933, Hitler in some ways showed considerable respect for both the technical expertise and the toughness of the German officer corps. When faced in summer 1934 with a choice between those radical elements of his own SA who wished to build their own army and the officer corps of the *Reichswehr*, Hitler chose the latter. Thus, while he built up the political strength and stability of his regime, Hitler willingly provided the armed forces with large sums of money to get on with the business of rearmament. He routinely attended the great military parades and demonstrations that increasingly became a part of life in the Third *Reich*, but did not interfere in the day to day discussions of doctrine and tactics.

But the harmony between Hitler and his generals began to unravel in the 1930s, with the breakdown coming in the diplomatic and political fields. Hitler had been less than satisfied with the reaction of his generals to the remilitarization of the Rhineland in 1936; the first crisis of early 1938 brought the forced resignation of Hitler's war minister and the commander in chief of the army.[22] The probable cause of the purge lay in a major disagreement that had occurred the previous November between Hitler, arguing for greater risk-taking in German foreign policy, and his generals and foreign minister who argued for a more cautious diplomatic approach. The argument between Hitler's radicalism and the conservatism of his opponents may also have involved differences over the tempo of rearmament, but it is unlikely that the two sides differed over the nature of the army's force structure.[23]

The next clash between Hitler and his officer corps occurred during summer 1938 over the Czech crisis. Here the nature of the argument between the Führer and his high command raises serious problems for the Blitzkrieg theory. Hitler pushed Germany and Europe towards a major military confrontation over Czechoslovakia; he argued that the major European powers would not intervene should Germany attack the Czech Republic and even if they did, the Wehrmacht possessed the strength to win a war in the east and in the west.[24] Considering the fact that Germany possessed only three panzer divisions, Hitler's analysis clearly put little weight on Germany's mechanized forces for achieving quick victories. On the other side, the military opposition, led by General Ludwig Beck, saw little chance for a German victory in 1938 and harbored serious doubts as to whether armored forces and the concept of a quick war were viable possibilities given the nature of European politics.[25] The generals hit closer to the mark. Only the abdication of the Western Powers at Munich prevented Hitler from launching the Wehrmacht against Czechoslovakia and causing a European war that would have ended his regime at a considerably earlier date than 1945.[26]

As suggested earlier, the German economy underwent a serious crisis in late 1938 as a result of the multiple strains under which it had been working. The problems that this caused for rearmament helped to precipitate the occupation of the remainder of Czechoslovakia in spring 1939. But while correctly estimating the weaknesses of both British and French governments, Hitler missed the fundamental changes in the popular attitudes of Western Europe. Thus, although he managed to isolate Poland even more thoroughly than Czechoslovakia, he could not get away with his "little" war but rather, unleashed World War II.[27]

Several factors led Hitler to risk war in September 1939, among them his feeling that Germany enjoyed decisive air superiority.[28] Ironically, there was almost no mention of superiority in armored strength and doctrine. Even so, the campaign in Poland proved so successful that many doubters in the higher levels of the officer corps, most notably Gerd von Rundstedt, were won over to the concept of armored, deep penetration operations. Several remained unconvinced, among them Franz Halder, chief of the general staff, who opposed the Ardennes strategy through March 1940.[29]

The declaration of war by the Western Powers in September 1939 caught Hitler by surprise and placed Germany in a dangerous strategic situation. Hitler faced the possibility of a blockade with little concern; he claimed that with supplies from Russia and the Balkans Germany need have no fear of an Allied blockade.[30] Hitler could not have been more wrong; in fact, Germany faced a catastrophic economic situation.[31] The advocates of the Blitzkrieg strategy theory suggest that because of the low output of munitions and weapons, the Germans deliberately initiated a long pause between the Polish campaign and the commencement of major operations in spring 1940. Hitler in fact exhibited little knowledge of the very strategy he supposedly designed. Instead of establishing a pause for Germany to build up stocks after the Polish campaign, the Führer pushed for an immediate offensive against the West. Why? Germany's economic situation appeared so weak that a pause seemed fraught with danger.[32] On the other hand, the generals opposed the immediate onset of hostilities against the West, not because of a desire to build up stocks, but rather because of their feeling that the army's tactical performance exhibited in Poland had not come up to their exacting standards.[33]

Over autumn and winter 1939-40, Hitler and his generals engaged in considerable squabbling about the operational conception of the coming campaign in the West. Only in March 1940 did the German high command agree to a great armored drive through the Ardennes. Even then some senior officers harbored serious doubts as to Guderian's chances of getting across the Meuse.[34] Hitler himself never fully understood the concept of deep penetration and exploitation and remained nervous throughout Guderian's drive to the coast. Rundstedt was equally worried and along with Hitler deserves a major share of the blame for the famous stop order that prevented Guderian's tanks from seizing Dunkirk and allowed a sizeable portion of the BEF to escape.[35]

The picture that we get of German grand strategy, then, is murky indeed. The conceptual framework of the Blitzkrieg strategy simply does not work. Hitler knew the long-range goals which he planned to achieve. How exactly he would reach those goals and destroy the European diplomatic and military balance of power, he had no clear idea. He did, however, recognize that realization of his ambition to destroy the European state system required a wholehearted commitment to armed conflict. Admittedly, he maneuvered with great skill from 1933 to 1938 as he overturned the diplomatic balance of power.[36] However, when he tipped his hand in March 1939 by seizing Czechoslovakia, he created the climate that resulted in the second great European war within twenty-five years. He possessed no clear conception of how Germany might win that war, only a steadfast belief that the Wehrmacht, army as well as air force, would beat Germany's opponents.

On the other hand the generals, no matter what their degree of German nationalism, balked at the prospect of war in 1938; in 1939 they supported the invasion of Poland both because events had proven the Führer right the year before and because the deal with Russia had removed the prospect of a war on two fronts. But when faced with the prospect of attacking France in autumn 1939, the generals hesitated to embark on another great European war, à la 1914. Some even resorted to conspiracy in an attempt to avoid another such catastrophe.[37] Nevertheless, the Germans put together a tactical and operational doctrine in 1939 based on trends within their military services reaching back to World War I. But that doctrine, which one might call Blitzkrieg tactics, did *not* evolve as the result of a grand strategic design.

In conclusion, a close examination of the evidence does not support the theory of a Blitzkrieg strategy; rather it points to an uncertain and unclear grand strategy in which the Germans put the military pieces together at the last moment, with serious doubts and considerable haste. The late 1930s suggest that statesmen and military leaders do indeed "see through a glass darkly," and that the interrelationships among strategy, national security policy, diplomacy, and political goals are often disorderly and chaotic. Any attempt to assign neat theoretical explanations to such processes, can only be made at the expense of historical reality.

NOTES

1.   For the theory of Blitzkrieg strategy, see Larry Addington, *The Blitzkrieg Era and the German General Staff*, vol. I (New Brunswick, 1971); Alan Milward, *The German Economy at War* (London, 1965); Burton Klein, *Germany's Economic Preparations for War* (Cambridge, Mass., 1959); and most recently F. H. Hinsley, *British Intelligence in the Second World War* (London, 1979).

2.   Wilhelm Deist, Manfred Messerschmidt, Hans-Erich Volkmann, and Wolfram Wette, *Das Deutsche Reich und der Zweite Weltkrieg*, vol. 1, *Ursachen und Voraussetzungen der deutschen Kriegspolitik* (Stuttgart, 1979), pp. 480-88; and Edward Homze, *Arming the Luftwaffe. The Reich Air Ministry and the German Aircraft Industry, 1919-1939* (Lincoln, 1976), p. 184.

3.   For a fuller discussion of these problems see Hans-Erich Volkmann, "Aushandeln und Aufrüstung in Deutchland, 1933 bis 1939," *Wirtschaft und Rüstung am Vorabend des Zweiten Weltkrieges*, Friedrich Forstmeier and Hans-Erich Volkmann, eds., (Düsseldorf, 1975), p. 85. One must note that the option of borrowing money to finance rearmament was extensively used by the Nazis. However, when internal sources of credit dried up in the mid-1930s, foreign bankers were most reluctant to extend Germany the long-term credits needed to finance the raw material imports on which rearmament depended. Thus, only through

exports could the Germans earn the foreign exchange needed to pay for raw material imports.

4.   Among many other incidents see Hitler's autumn 1938 demand that the Luftwaffe's strength be quadrupled. Homze, *Arming the Luftwaffe*, pp. 223-24.

5.   International Military Tribunal (IMT), *Trial of Major War Criminals* (TMWC) vol. XXXII, Doc. No. 3575, p. 413.

6.   Jost Düffler, *Weimar, Hitler und die Marine, Reichspolitik und Flottenbau 1920-1939* (Düsseldorf, 1973), p. 504.

7.   Klein, *Germany's Economic Preparations for War*, pp. 19-21, 51, 79.

8.   Anja E. Bagel-Bohlan, *Hitlers industrielle Kriegsvorbereitungen, 1936-1939* (Koblenz, 1975), pp. 17, 72.

9.   MacGregor Knox, *Mussolini Unleashed 1939-1941* (Cambridge, 1982), Appendix 2.

10.   See Georg Thomas, *Geschichte der deutschen Wehr-und Rüstungswirtschaft, 1918-1943/5* (Boppard am Rhein, 1966).

11.   Williamson Murray, *The Change in the European Balance of Power, 1938-1939* (Princeton, 1984), chapter 5.

12.   Edward W. Bennett, *German Rearmament and the West, 1932-1933* (Princeton, 1979), p. 324.

13.   Friedrich Hossbach, *Zwischen Wehrmacht und Hitler, 1934-1938* (Hannover, 1949), p. 39.

14.   For further elaboration on this point, see Murray, *The Change in the European Balance of Power*, chapter 1.

15.   For the fact that the French were close to the Germans in terms of numbers of weapons see R. H. S. Stolfi, "Equipment for Victory in France in 1940," *History* (February 1970).

16.   Among other articles see Major Bartz, "Kriegsflugzeuge; ihre Aufgaben und Leistung," in the main German military journal, *Militärwissenchaftiche Rundschau* (1936), p. 210; and Major Herhudt von Rohden, "Betrachtungen über den Luftkrieg," Part I (1937), p. 198. For a more detailed examination of this problem see my article "The Luftwaffe before the Second World War: a Mission, a Strategy?," in *Journal of Strategic Studies* (September 1981).

17.   BA/MA RL 7/42, RL 7/43, Luftflottenkommando 2, Führungsabteilung, Nr. 7093/39, 13.5.39, *Schlussbesprechung des Planspieles 1939*.

18.   See in particular *Documents on German Foreign Policy* (DGFP), Series D, vol. VII, Doc. 192, 22.8.39.

19.   See the figures in the "Luftwaffe Strength and Serviceability Tables, August 1938-April 1945," Air Historical Branch Translations No. VII/107, for the strength of the Luftwaffe's close support forces. Aircraft such as the He 111, Do

17, and Ju 88 were admittedly not satisfactory "strategic" bombers, but then the Luftwaffe had never regarded them as such. They were an interim bridge in bomber production to fill the gap between the Ju 52 and the He 177—an aircraft which, no matter what its technical deficiencies, the Germans regarded as a serious "strategic" bomber. It is also worth noting that in the 1930s the Germans did believe that within the distances of Central Europe the He 111, Do 17 and Ju 88 could perform as "strategic" bombers. So did the other airmen of the world.

20. "Lehren aus dem Feldzug in Spanien, Einsatz von Schlachtfliegern," aus einer Studie der 8. Abt. des Generalstabes aus dem Jahre 1944; Hans Hennig Freiherr von Beust, *Die deutsche Luftwaffe im spanishchen Krieg*, 2.10.56, p. 162, Albert Simpson Historical Research Center: Karlsruhe Collection (ASHRC:KC) KC 113.302.

21. *DGFP*, Series D, vol. VII, Doc. 307, 26.8.39.

22. For a fuller discussion of the crisis, see Harold Deutsch, *Hitler and his Generals* (Minneapolis, 1974).

23. For an excellent description of the November 1937 meeting and its implications see Alan Bullock, "Hitler and the Origins of the Second World War," Hans W. Gatzke (ed.), *European Diplomacy Between Two Wars* (Chicago, 1972).

24. See Murray, *The Change in the European Balance of Power*.

25. BA/MA N28/3, Beck Nachlass, "Betrachtungen zur gegenwärtigen mil. politischen Lage," 5.5.38, and "Bemerkungen zu den Ausführung des Führers am 28.5.38," 29.5.38.

26. See Williamson Murray, "Munich 1938: the Military Confrontation," *Journal of Strategic Studies* (December 1979).

27. Christopher Thorne, *The Approach of War, 1938-1939* (New York, 1969).

28. *DGFP*, Series D, vol. VII, Doc. 192, 22.8.39.

29. Heinz Guderian, *Panzer Leader* (New York, 1952), pp. 90-92.

30. IMT, *TMWC*, vol. XXVI, Doc. 798 ps.

31. Schlesisches Institut für Wirtschafts-und Konjunkturforschung, "Zahlen des deutschen Aussenhandel seit Kriegsbeginn (August 1940)," National Archives and Records Service (NARS) T-84/195.

32. OKW files: "Denkschrift und Richtlinien über die Führung des Krieges im Western," Berlin, 9.10.39, NARS T-77/775.

33. See my article, "The German Response to Victory in Poland: a Case Study in Professionalism," *Armed Forces and Society* (Winter 1980).

34. Guderian, *Panzer Leader*, pp. 90-92.

35. The most thorough analysis in English is in Telford Taylor, *The March of Conquest* (New York, 1959), pp. 255-63.

36.   See in particular Gerhard Weinberg, *The Foreign Policy of Hitler's Germany, 1933-1936* (Chicago, 1970).

37.   Harold Deutsch, *The Conspiracy Against Hitler in the Twilight War* (Minneapolis, 1968).

CHAPTER TEN

THE GERMAN RESPONSE
TO VICTORY IN POLAND:
A CASE STUDY IN PROFESSIONALISM

Throughout history, military organizations have attempted to learn from experience. However, they have tended to extract from their experiences only that data which supports previously held notions. In many cases, existing doctrine creates a barrier to adaptation and improvement, with military historians often ignoring both the process by which doctrine and battlefield experiences interact, and the ways in which armies adapt or do not adapt to changing conditions. By concentrating on operational history, the straightforward account of campaigns and generalship, historians have not paid sufficient attention to the hows and whys. As a result, training, doctrine, and the process of learning from experience play almost no role in descriptive military history. Yet these factors may well spell the difference between success or failure on the battlefield.

This chapter addresses an area typically of little interest to most military historians: the process by which a military organization, in this case the German army, prepared to overcome the tactical problems it faced in waging the great offensive against the West in the spring of

1940. It will examine those long, dull, tedious processes of training and changes in doctrine, for, as Clausewitz points out in *On War*:

> There are only two sources of this [military] spirit . . . The first is a series of victorious wars; the second, frequent exertions [training] of the army to the utmost limits of its strength. Nothing else will show a soldier the full extent of his capacities. The more a general is accustomed to place heavy demands on his soldiers, the more he can depend on their response. A soldier is just as proud of the hardships he has overcome as of the dangers he has faced. In short, the seed will grow only in the soil of constant activity and exertion.[1]

In recent literature, scholars have rejected the legend that the Germans enjoyed immense material and numerical superiority in the 1940 campaign.[2] Historians instead have centered their attention on strategic planning and the actual conduct of operations by the opposing sides in their explanations for the overwhelming nature of the German victory.[3] This chapter will offer another possibility for that victory by examining the evolution of German training and doctrine in the period of the "phony war." This two-faceted process began with a close examination of the lessons of the Polish campaign and their application to German doctrine and training, followed by a rigorous training program that brought the army's reserve and *Landwehr* units, all varying widely in their composition, training, and doctrine, up to the standards of regular units. What must be emphasized is that the German army in its "lessons learned analysis" of the Polish campaign did not use its studies to support existing doctrine. Rather it used its after-action reports to improve doctrine and military standards throughout the army.

On the surface, the Polish campaign of September 1939 proved a smashing success for the German army. In less than one month, the German army completely shattered an army of forty divisions and one million men, and two-thirds of Poland lay in German hands. In every sense this campaign was an outstanding operational success. Yet, the OKH (*Oberkommando des Heeres*) judged this operational success as insufficient and inadequate. In general, the after-action reports (*Erfahrung-sberichte*) of the German army for the period 1938 to 1940 reflected an unusual pessimism. The higher the headquarters, the more demanding and dissatisfied the commanders became with operational performance.

During this same period of time the entire German system involved a great deal of trust and honesty between different levels of command. German officers in command positions were not afraid to express their belief that their units were deficient when circumstances justified such comments. The *Anschluss* is a particularly good example of this process, as the occupation of Austria highlighted serious weaknesses within the army and the mobilization system. After-action reports from the battalion to the army level became more and more critical of troop performance, training, discipline, and doctrine the higher the level of command.[4] This willingness to be self-critical was one of the major factors enabling the German army to perform at consistently high levels throughout World War II.

A report on the Polish campaign by the Twelfth Army underlined not only the value of the training that its officers and NCOs received, but also the close connection between that training and the experiences gained in the occupation of the Sudetenland and Prague during the previous year.[5] Similarly, an after-action report of the 6th Panzer Division on the campaign in the west in 1940 stressed the role that the Polish campaign had played in improving armor doctrine.[6] At the conclusion of the victory over Poland, the OKH pointed out to subordinate commands that "in the interest of the whole army [it needed] to collect as soon as possible the combat experiences in both the tactical and technical spheres," to disseminate these experiences widely among the troops, and to use them as the basis for training the replacement army.[7]

In its preliminary report on the war against Poland, the OKH indicated that while the Polish campaign "fully confirmed our tactical principles," a number of significant weaknesses had shown up. Moreover, the OKH felt that a number of important principles needed reemphasis. First, effective combat leadership, no matter what the level, must be from the front rather than from the rear. Secondly, frontline commanders tended to exaggerate their losses, the enemy's strength, and the difficulty of terrain in their combat reports and, thus, needed work on sending accurate, terse combat information. Finally, German troops had proved inadequately prepared to carry out reconnaissance or security missions.

In terms of offensive tactics, the OKH judged infantry fire and cooperation between weapons and branches as unsatisfactory. The infantry failed to receive the necessary support from their heavy weapons and the artillery. The transition from offense to defense did not work well, and German troops often formed linear defensive systems rather than the defense in depth demanded by German doctrine. Finally, although the OKH did pay tribute to the Luftwaffe's contributions in the Polish campaign, it closed on the warning note that this air superiority caused the troops to become careless in camouflaging their positions and movements. Such carelessness in the west might have disastrous implications before air superiority was won.[8]

Reports on the performance of panzer and motorized divisions followed similar lines. Guderian's war diary reported that the performance of his corps fully lived up to peacetime expectations of what armored and mechanized forces could achieve in war.[9] However, Guderian does admit in his memoirs that on at least one occasion he intervened actively in company level operations in order to keep the momentum of his advance going.[10]

Although panzer operations confirmed operational expectations, serious tactical weaknesses did occur. Cooperation between infantry and armor had not always been successful. In some cases panzer commanders divided up their tank assets rather than use their armor as a unified, powerful striking force. The 7th Panzer Division (formerly 2nd Light) reported that attacks must be carefully prepared and executed, while avoiding a too methodical approach that would rob subordinate commanders of initiative.[11]

A number of after-action reports condemned march discipline and traffic regulation for both infantry and motorized forces as completely unsatisfactory.[12] Light divisions, established in 1935 as a sop to the cavalry, performed adequately in Poland, but German reports stressed that conditions in Poland might not be the same in the west.[13] Before the war Guderian argued that the light divisions could not match the staying power of panzer divisions and that the Luftwaffe could better perform accurate and economical long-range reconnaissance.[14] The 2nd Light Division's after-action report confirmed Guderian's criticism.[15] As a result of these experiences the OKH decided to convert light divisions into panzer divisions. In the final analysis, the Polish campaign served not only to substantiate the opinions of tactical innovators like Guderian, but also to win over a substantial number of those who had opposed armored tactics before the war, such as Rundstedt and Rommel.[16]

The infantry and mountain divisions that made up nearly 80 percent of German forces learned similar lessons. In late November the OKH reported that the Polish campaign confirmed the capabilities of German infantry, its organization, weapons, and doctrine. However, requests from the regular infantry establishment for more motorization, tanks, and armored artillery pieces, went unanswered because of a reluctance either to divide up valuable assets or to violate the principle of concentration (*Schwerpunktsbildung*).[17] The 8th Infantry Division emphasized at the beginning of its *Erfahrungsberichte* that "infantry tactics and training principles as well as German organization and weapons had completely proven themselves" in combat.

But difficulties did emerge. German infantry did not meet expectations in either night fighting or in combat in heavy terrain. Troops tended to halt and fire when running into enemy units thus forcing the enemy to ground and failing to push attacks home.[18] Unlike some units, Twelfth Army reported that cooperation between infantry and artillery was outstanding (*ausgezeichnet*). However, battalions formed from reservists had not done well, suffering numerous casualties from exhaustion during extended marches. Moreover, reservists lacked the necessary training and cohesion to stand up to the demands of German elastic defensive tactics. The report also added that the Poles had been far better prepared for night fighting than the Germans.[19]

Immediately after the conclusion of the Polish campaign, the OKH amassed a considerable amount of critical, detailed after-action reports from the highest level of command down to regimental level. German commanders showed little fear of repercussions from the critical comments and evaluations of their unit's performance. In fact, General Walter von Brauchitsch, commander in chief of the army, underlined his expectation that subordinate commanders would pass critical evaluations up the chain of command. In October 1939, he established a monthly evaluation report for divisional and corps commanders to indicate the level of combat effectiveness of their units in order to avoid the mistakes of the German high command in World War I in overestimating the fighting ability and capabilities of frontline units.[20] These reports and after-action critiques formed the basis of the entire German training program over the winter of 1939-1940.

The Germans felt that the Polish campaign confirmed their doctrine and training. Those deficiencies noted, however, made the OKH dubious about the possibility of launching the Wehrmacht against the west before the coming spring. This attitude led to a blowup between Hitler and the OKH in the fall of 1939 with Hitler demanding that the army launch an immediate offensive in the west, while the generals argued that the army was not ready.[21] Both sides had considerable basis for argument. Hitler focused on an economic situation that looked cata- strophic.[22] The generals argued that even regular army units displayed weaknesses in Poland requiring considerable time for correction.[23]

But looking at the German army as a whole rather than just those units that participated in the attack on Poland considerably strengthened the generals' argument. By 7 September 1939, the OKH had managed to deploy thirty-five divisions in the west; only eleven represented active duty, regular infantry divisions. Six divisions came from the 1906-1912 conscription classes that had received no military training because of the Treaty of Versailles. These individuals had only recently received call-up notices and had been on active duty for less than two months. Nine divisions consisted almost entirely of *Landwehr* personnel who had served in World War I but had received almost no training in the intervening period. Two divisions consisted of border patrol and security units, also made up in part by *Landwehr* personnel. Finally, seven reserve divisions drew their strength from reservists who had only recently completed military service.[24]

While the regular army units in the west and the reserve divisions in most cases came up to the standards of the general staff, the other formations faced serious shortages in terms of personnel, equipment, and training.[25] To further complicate the situation, throughout fall and winter of 1939-40 the Germans carried out a rapid expansion of the army, as levies of untrained Germans received call-up notices. The es- tablishment of new divisions led to the transfer of experienced officers and NCOs from regular units. Army Group A in particular complained about the effects of this practice.[26] Finally, the motorized and mechanized forces went through a major reorganization program as mentioned above.

It seems clear why the German army high command so opposed a major offensive in the west before the spring of 1940. Not only did it want the regular units to absorb the lessons of the Polish campaign, but it also wanted to correct the vast difference between the training and readiness of the more inexperienced divisions and the regular army. The process by which the German army managed to bring these reformed divisions up to the standards of the regular army played a major role in the German victory in the west in the spring of 1940.[27]

On 13 October, the OKH released a memorandum to subordinate commands on "The Training of the Field Army."[28] This command memorandum formed the basis for the training program of the German army over the coming six months and received wide dissemination throughout the army structure.[29] The OKH pointed out that:

> the conclusion of the Polish campaign and the quiet in the west will give the field army the possibility of perfecting its performance, discipline, and coherence. . . . The exclusive goal of each training exercise is the insertion of troops in battle. Beside weapons and battle training the education of the soldier stands in the foreground. Troops are to be hardened and prepared to meet the highest demands of war, especially against an enemy trained and equipped with modern weapons.

In specific terms, the OKH ordered that troops of the first, second, fourth, and fifth waves of mobilization be "prepared to attack fortifications; to exploit success; to defend themselves against enemy attacks and tanks." The OKH strongly emphasized that the experiences of Poland be tied directly to the training: reconnaissance and security, march discipline, infantry fire discipline, cooperation between weapons, offensive, and defensive tactics at twilight and during the night, and conversion from an offensive to a defensive posture.

The second section of the report emphasized that training must stress leadership, especially the combat leadership of the German officer corps. Reserve officers and NCOs should similarly meet the standards of the peacetime army, in part so that "the position of the NCO as leader, trainer and educator would not sink as it had in the last years of the First World War." And finally, the OKH underlined that "discipline was the basis of victory" and that "all ranks must be clear that the maintenance of discipline and order in form as well as in military courtesy would have to be implacably maintained."

The general control of the training program by the OKH involved the scheduling of divisions for intensive work on the military reservations such as Grafenwöhr and Ohrdruf.[30] Moreover, the OKH took control of scheduling officers and NCOs to attend training schools. Most of the staff from the infantry training school at Doberitz transferred to active units on the western front while the OKH brought in officers and NCOs with combat experience either in the west or in Poland.[31] Moreover, the general staff's training section laid out a detailed set of standards and objectives for the training schedules of artillery and infantry schools within the army groups. A Sixth Army circular stressed that the current training program aimed to improve the standards of officers and NCOs and to train new units in offensive tactics.[32]

OKH interest in the training program extended to the point of actually laying out the 12-day schedules of training for company and platoon commanders.[33] Moreover, the OKH required lower headquarters to submit *Erfahrungsberichte* on the school sessions that their personnel attended.[34] The reports from these special training programs contain several interesting points on the composition and level of training within the army. For the first training session within First Army, the average age of the sixty-two officers attending the course for company commanders was thirty-seven, while over 50 percent of the officers were forty or over. Of these officers, 16 percent were regular officers while the remainder were reserve or *Landwehr* personnel.[35] A report from the anti-tank school at Baumholder indicated that a number of officers in attendance had never before seen a German tank.[36]

The training programs within the German divisions aimed at first building up the capabilities of the individual soldier and then establishing the capabilities of the platoons and companies. Once small-unit training reached satisfactory levels, then larger scale battalion and regimental exercises could be conducted. Inadequate training facilities in the west complicated training programs and both battalion level maneuvers and large-scale training often met with delays until divisions could move to regular training areas.[37] The internal reports and critiques of training programs within divisions indicate a carefully detailed and methodical approach. Above all, division staffs urged that training be expedient. The 88th Division suggested that at the end of every exercise the instructor should ask himself whether "his personnel had learned something or not?"[38] The Germans viewed realism in training as being of inestimable importance, and one training officer considered exercises that did not give a realistic impression of the enemy and of combat conditions as useless.[39]

Like the OKH, division commanders consistently demanded that discipline and appearance be rigidly maintained as one of the key elements in unit morale.[40] As the staff of the 44th Division commented: "The sharpest discipline of units and individual soldiers is the basic principle."[41] This same report captured one of the key elements in the success of the German army in the coming war with the criticism that lower ranking officers and NCOs seemed unwilling to show sufficient initiative. Thus, senior commanders should encourage their subordinates to take initiative, while at the same time not ignoring their stated orders. This tension between the emphasis on initiative and the emphasis on obedience contributed heavily to German successes early in the war. The army in effect demanded that its officers use their judgment. An officer who failed through an unwillingness to seize the initiative deserved as much condemnation as one who failed by ignoring his orders.

The experience of the 208th Division as it worked up to combat ready status over the winter of 1939-1940 is representative of reserve division training. The division formed up in the summer of 1939, shortly before the outbreak of war. Initially it contained a substantial percentage of *Landwehr* personnel and in October it moved to the town of Sobotka in the military district of Posen. There it began an arduous program to turn itself into a first-class infantry division. The training program aimed to exercise the troops while emphasizing weapons training, the duties of the German soldier, technical training, and the utilization of terrain. The division staff gave special emphasis to upgrading junior officers and NCOs. The initial training schedule called for two weeks of individual training, one week for squad training, one week for platoon training, one and one-half weeks for company training, and one and one-half weeks for larger unit training.[42]

The training program succeeded in weeding out a number of reservists and *Landwehr* personnel who could not meet the standards of the division, while a number of untrained replacements arrived to fill any gaps. Urged on by its corps commander, the 208th attempted to integrate these replacements as rapidly as possible.[43] In February it reported to corps headquarters that it had ceased large-unit training in order to devote all training time to the replacements.[44] By March the division returned to large-unit training that emphasized offensive spirit and movement, *"schnell vorwärts."*[45] The division commander reported in March that the February training had successfully integrated the replacements at an acceptable level of competence. Moreover, he reported, in only three months the new troops had achieved a state of combat readiness that fully corresponded to the demands of the army.[46]

The brief training reports and plans only hint at the rigor with which the German army carried out its program to bring the German soldier up to its standards.[47] The training aimed to push officers and men to the limits, both in terms of physical conditioning and mental stamina. By demanding that troops be exercised to the breaking point, the German army sought to eliminate those who would break in combat. As the training program for the 10th Panzer Division suggested; the troops must be taught that "the removal of bad elements lies in their own interest."[48]

Above all the training aimed to foster an offensive spirit among German troops preparing for the coming push in the west. *Generaloberst* Fedor von Bock in a circular to his Army Group B at the end of April 1940, still voiced concern over the level of offensive spirit reached by his troops and commanders:

> In many exercises recently, particularly at the battalion and regimental level, an inclination to caution and circumspection has appeared. Therein lies the danger that on one side German leadership will pass up opportunities to seize favorable situations . . . While on the other hand the enemy will be allowed time to recognize our intention. . . [O]nce a commander has decided to attack, so must everything that he orders be established that the eyes, heart and senses of the troops be directed to the front.[49]

In retrospect, Bock's fears appear exaggerated. Throughout its units, whether they be regular, reserve, panzer, or line infantry, the German army had reached a state of combat readiness far superior to that of its opponents. Moreover, the regular and reserve line infantry divisions played a crucial role in making possible the devastating victory of May 1940. The brilliant Manstein plan for an armored thrust through the Ardennes has justifiably received much attention from historians. However, Bock's Army Group B played an equally vital role in the victory.

At the initiation of the offensive, Bock's forces, nearly entirely infantry divisions, diverted Allied attention to northern Belgium and Holland. Once the extent of the German breakthrough became clear, the hammering pressure of Bock's advance made it impossible for Allied forces to disengage in order to meet the threat from the south. Much of the German success in 1940 was thus the direct result of training and preparation.

NOTES

1.   Reprinted with permission of the publisher, Seven Locks Press, Cabin John, Maryland.

2.   Carl von Clausewitz, *On War*, ed. and trans. by Michael Howard and Peter Paret (Princeton, 1976), p. 179.

3.   See in particular R.H.S. Stolfi, "Equipment for Victory in France in 1940," *History* (February 1970).

4.   The German army and the campaign of May 1940 have of course received much attention. Works on the German army in the prewar period, however, for the most part discuss relations between the army and the Nazi party; see in particular Klaus-Jürgen Müller, *Das Heer und Hitler* (Stuttgart, 1966); Robert J. O'Neill, *The German Army and the Nazi Party* (London, 1966); and Manfred Messerschmidt, *Die Wehrmacht im NS-Staat* (Hamburg, 1969); works on the 1940 campaign stress the strategic development and operational history of events. See in particular: Alistair Horne, *To Lose A Battle* (Boston, 1969); and Telford Taylor, *The March of Conquest* (New York, 1958).

5.   Among others, see in particular: Heeresgruppenkommando 3., 18.7.38, "Der Einsatz der 8 Armee in März 1938 zur Wiedervereinigung Österreichs mit dem deutschen Reich," National Archives and Records Service (NARS) T-79/14/447; and Generalkommando XIII A.K., "Erfahrungsbericht über den Einsatz Österreichs März/April 1939," 6.5.38. NARS T-314/525/000319. German experience reached back before the Anschluss. Von Thoma underlined that nothing new had been learned in Poland that had not already been indicated in Spain. Panzer Regiment 3, 20.1.40 "Erfahrungen aus dem poln. Feldzug," NARS T-78/

379/6344436.

5.   Armeeoberkommando 12, Betr.: Erfahrungsbericht, 25.10.39, NARS T-315/671/000890.

6.   6. Pz. Div., 18.7.40, "Erfahrungsbericht der 6. Pz. Div., Feldzug Frankreich," NARS T-311/49/7061274.

7.   OKH, Berlin, Okt 7, 1939, Betr.: "Erfahrungsbericht bei der Operationen im Osten," NARS T-315/435/00491.

8.   OKH, Gen StdH O Qu I/Ausb. Abt. (Ia), 15.10.39, Taktische Erfahrungen im polnishchen Feldzug, NARS T-315/436/00462; for the same report issued to the 16th Division see NARS T- 315/671/000909.

9.   KTB XIX A.K., Feldzug in Polen, 1.9.39—25.9.39, p. 211.

10.   Heinz Guderian, *Panzer Leader* (New York, 1957), p. 51.

11.   7. Division, 31.10.39, "Auszug aus den Erfahrungsberichten des XVI und XIX Armeekorps sowie des A.O.K. 1O," NARS T-315/436/000467; see also Korpskommando XV A.K., 3.10.39, "Erfahrungen auf taktischen Gebiet," p. 3, NARS T-314/550/000297; KTB XV A.K., Der Feldzug in Polen, p. 35, NARS T-314/550/000256.

12.   See among others, 16. Div., 31.10.39, NARS T-315/671/001042: Erfahrungsbericht des XV A.K. bei den Polenfeldzug; Anlage 1: "Erfahrungen beim Marsch motorisierter Verbände während des Feldzuges in Polen," NARS T-312/37/7545415.

13.   Korpskommando XV A.K., 3.10.39, "Erfahrungsbericht über den Feldzug in Polen," NARS T-314/550/000297.

14.   Heinz Guderian, "Schnelle Truppen einst und jetzt," *Militärwissenschaftliche Rundschau*, Heft 2 (1939), p. 241.

15.   7. Panzer Division (2. Leichte Division), 19.10.39, Betr: "Erfahrungen bei den Operationen im Osten," NARS T-315/436/000480.

16.   See Murray, *The Change in the European Balance of Power*, chapter 1.

17.   OKH. 28.11.1939, "Auswertung der Erfahrungen bei den Inf.- u. Geb. Division in Polen Feldzug," NARS T-312/234/7788373.

18.   8. Division, 19.10.39, Erfahrungsbericht, NARS T-314/372/000189.

19.   Armeeoberkommando 12, Erfahrungsbericht, 25.10.39, NARS T-315/671/000890.

20.   Der Oberbefehlshaber des Heeres, 24.10.39, Zustandsberichte, NARS T-315/1025/357.

21.   See Harold C. Deutsch, *The Conspiracy Against Hitler in the Twilight War* (Minneapolis, 1968).

22.   See Murray, *The Change in the European Balance of Power*, chapter 5.

23.   For two examinations of the weakness of troop training and the inade-

quacy of the field army see Heeresgruppe A, Ob, 31.10.39, An den ObdH, NARS T-311/236/884; and Heeresgruppe A, Abschrift von Abschrift, Besprechung am 11.11.39, NARS T-311/236/853. Some officers overstated the weaknesses that showed up. This led the infantry desk at OKH to issue a rebuttal to the charge that German infantry in Poland had not performed as well as German infantry in 1914. General der Infanterie beim Ob.d.H., 30.11.39, "Der Angriffsgeist der Infanterie (Erfahrungen des Ostkrieges)," NARS T-311/47/7057941.

24.   The breakdown of German strength on the western front at this time came from the following documents 1. A.O.K., "Wir halten-wir stürmen: Beton und Stahl: Kampf und Sieg der 1. Armee im Westwall und Maginotlinie," NARS T-312/1/7500002; Oberkommando Heeresgruppe C, "Verstarkung der 5. Armee," NARS T-312/36/7544294; Anlage 1 zu H.Gru.Kdo.C, Ia Nr 2580/39, 6.9.39, "Aufstellung und Einsatz der Armee-Abt. A," NARS T-311/38/7646927; "Anlage 1 zu A.O.K. 5, Ia Nr 200/39, 21.8.39," NARS T-311/125/ 7168743; A.O.K. 5., Abt., Ia, Nr11O/39, 7.9.39, Betr: Armee Abteilung A, NARS T- 311/40/ 7050454; and U.S. Army, "Order of Battle of the German Army" (Washington, D.C., 1945).

25.   For an interesting description of the inadequacies of these divisions, see Siegfried Westphal, *The German Army in the West* (London, 1957).

26.   Heeresgruppe A im Westen, 1939/1940; 1 Teil: Vorbereitungszeit, 25.10.39-9.5.40, p. 5, NARS T-311/236/ 705.

27.   It is also worth noting that the Waffen SS also expanded from its brigade formations to a force of motorized infantry divisions. For a description of the working up of one of these divisions, SS Totenkopf, see the outstanding work by Charles W. Sydnor, Jr., *Soldiers of Destruction* (Princeton, 1977).

28.   Des Oberbefehlshaber des Heeres, Gen St d H/Ausb. (Ia), Nr. 400/39g, 13.10.39, Betr.: "Ausbildung des Feldheeres," NARS T-312/234/7787781.

29.   See: 16. Division, "Richtlinien für die Erziehung und Ausbildung," 24.10.39, NARS T-315/671/000868; and Brauchitsch's circular from 12.12.39, Gen Std H/Ausb. Abt. (Ia), Nr. 800/39g, Betr.: "Ausbildung des Feldheeres," NARS T-312/234/7788296 or T-312/752/8396808.

30.   Oberkommando des Heeres, Gen St d H/Ausb. Abt. (Ia), Nr. 135/40g. 19.1.40.

31.   Oberkommando des Heeres. Betr.: "Austausch von Ausbildungspersonal des Infanterieschule," 14.12.39, NARS T- 311/47/7057948.

32.   Armee-Oberkommando 6, 14.12.39, Betr.: "Ausbildung der Feldheeres," NARS T-312/752/8396799.

33.   Ob d H Gen St d H/Ausb. Abt. (Ia) Nr. 900/39g. II Ang, 13.12.39.

"Stoffplan für einen Kompanieführer-und Zugführerlehrgang der Infanterie und Stoffplan für einen Lehrgang der Unterführer der schweren Infanteriewaffen," NARS T-312/752/8396772; and OKH Gen. St. d. H.- Gen. d. Art., Ia/Nr. 40/40, 4.1.40. "Lehrplan für Batterie-führer-Lehrgänge bei den H. Gr," NARS T-311/47/7057905.

34.   Heeresgruppenkommando C, An OKH, Gen. St. d. H./Ausb. Abt, Nr. 1300/40; A.O.K. 1, Ia/Art. Nr. 174/40, 21.2.40. An Oberkommando Heeresgruppe C., NARS T-311/47/7057900-904.

35.   Inf. Lehrbataillon A.O.K. 1, 14.2.40, "Erfahrungsbericht über 1. Kp. Führer-Lehrgang," NARS T-311/47/7057908. While most reports on senior company grade officers remained satisfactory, the 211th Division used the occasion of a visit by the commander of Army Group B to suggest that the average age of the officer corps should be substantially lowered. 211. Div., Abt. Ia, 16.12.39, Betr.: "Besuch des Herrn Oberbefehlshabers der Heeresgruppe B." NARS T-314/238/000014.

36.   Panzer-Abwehr-Lehrgang für Kompanieführer, Tr. Übung Pl. Baumholder, 14.2.40, Erfahrungsbericht, An A.O.K.5, Abt. Ia., NARS T-311/47/7057910.

37.   211. Division, Abt Ia, Betr.: "Gelände für die Verbandsausbildung," 10.12.39, NARS T-314/238/029.

38.   88. Inf. Div., 18.1.40, Abt. Ia. "Bemerkungen für Ausbildung und Erziehung" (Nr. 1-Nr. 3), NARS T-315/171/706/712.

39.   Ibid.

40.   Among others see 87. Inf. Div., Abt. Ia. Nr. 117/40, 14.3.40, "Richtlinien für die Ausbildung," NARS T-315/1139/389; and 211. Division, Abt. Ia Nr. 344/39, 27.10.39, Betr.: "Ausbildung," NARS T-314/238/65.

41.   4. Division, Abt. Ia. Betr.: "Bemerkungen zu der Übersetzung am 19.4.40." 20.4.40, NARS T-315/910/467.

42.   208. Infanterie Division, Abt. Ia. Betr.: "Ausbildung während des Einsatzes im Mil. Bezirk Posen," 10.10.39, NARS T-315/1609/464.

43.   Korps Kommando XXXIII. Oberbefehlshaber, 10.1.40, 68/4O, NARS T-315/1609/549.

44.   208. Infanterie Division, Abt Ia, 6.2.40, Betr.: "Vorlage von Dienstplänen und Ausbildungsrichtlinien." NARS T-315/ 1609/584.

45.   208. I. D. Abt Ia, Tgb. Nr. 526/40, 8.3.40, "Allgemeine Richtlinien für die Ausbildung," NARS T-315/1609/608.

46.   208. I. D., Betr.: "Beurteilung des eingetroffenen Ersatzes," März 1940, NARS T-315/1609/643.

47.   For graphic illustrations of what training involved, consult Guy Sajer, *The Forgotten Soldier* (New York, 1971); and Hermann Teske, *Bewegungskrieg*

(Heidelberg, 1955).

48.   Tenth Panzer Division, Abt Ia Nr. 141/39, Betr.: "Ausbildung," 31.10.39, NARS T-315/558/812.

49.   Heeresgruppe B, Ia Nr. 2211/4O, 28.4.194O, "Bemerkungen zu den Truppenubungen in Frühjahr, 1940," NARS T-312/752/8396741.

INDEX